Beyond the Paranormal:

Learning from the Past at Haunted Locations

John G. Sabol, Jr.
C.A.S.P.E.R. Research Center

Also by John Sabol

Ghost Excavator (2007)

Ghost Culture (2007)

Gettysburg Unearthed (2007)

Battlefield Hauntscape (2008)

The Anthracite Coal Region (2008)

The Politics of Presence (2008)

Bodies of Substance, Fragments of Memory (2009)

Phantom Gettysburg (2009)

*Digging Deep: An Archaeologist Unearths
a Haunted Life* (2009)

The Re-Haunting(s) of Gettysburg (2010)

*The Haunted Theatre: Digging Deeper
Into A Landscape of Ruins* (2011)

Ghost Culture Too (2012)

*Burnside Bridge: Expanding the Contemporary Reality of
Past Interactive Interactions* (2013)

Beyond the Paranormal:

Learning from the Past at Haunted Locations

Ghost Excavator Books, Inc™©

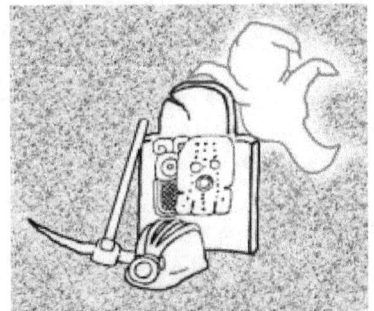

Brunswick, Maryland, USA

"It is the dead who make the longest demands on the living".

- **Sophocles, <u>Antigone</u>**

"The terrible gift that the dead make to the living is that of sight, which is to say foreknowledge; in return they demand memory, which is to say acknowledgement"

- **L. Sante, <u>Evidence</u> (1992)**

This "acknowledgement" must become the legacy of ghost research. This is a lesson we must learn through <u>constructive</u> fieldwork. It is a "heritage" we must preserve to legitimize that research……..

"Paranormal"

What is "paranormal", and what is "beyond" it? "Paranormal is something that cannot be explained at this time by science, or can be explained with certain revisions to current scientific principles. Paranormal is also something that is not compatible with normal perceptions, expectations, and cultural beliefs. Though not a thorough definition of "paranormal", it is sufficient to analysis the most basic tenets of this concept. Let's look at some key words:

- "Science": What "science" are we talking about, and using as a reference? There are many different sciences, not one exclusive scientific methodology, and each has their own set of operational procedures;
- "Normal": What is "normal"? Define it. What standards are used? Is today's "normal" yesterday's "normal"? ;
- "Expectations": If I am a field investigator, and researching an "unknown", then I am used to expect the "unexpected". My "expectations" are different from investigating already known qualities;
- "Cultural Beliefs": Whose? Are these the cultural beliefs of today's society, or those we are attempting to study?

We must learn to ask ourselves these questions, and not always question the "ghosts" ("Is anyone here with us tonight") because the questioning of "ghosts" is "paranormal". It assumes that they do <u>not</u> exist, instead of granting them presence, and exploring <u>their</u> lives, not their death. "Beyond" the "paranormal" is going back to the past, and exploring what still remains of that past in the present. It is <u>not</u> something <u>beyond</u> our capacity to learn new things about that past.....

Acknowledgements

I would like to thank the following individuals for their participation in these "ghost excavations" as a learning process. Without their willingness to think and act "outside the box" of the "paranormal", these projects would have not been completed. And they would not have been "performed" in a manner consistent with an alternative view to traditional "ghost hunting".

The Knickerbocker "Excavation":

Margaret Byl	James Castle
Mari Chastain	Michelle Desorchers
Guy Fazio	James McCann
Adriana Nicolleta F	Bert Richards
Herman (Mike) Stevenson	Don McDaniel

and Mary Becker

The Burnside Bridge "Excavation":

Renee Cannon	James Castle
Gary French	Cathy "Cat" Gasch
Linda Good	Rie Sadler
Jim Jones	Matthew James
Timothy Manville	Sally Kirchoff
Joseph Ritter	Bert Richards
Craig Rupp	Natalie Rose
Mike Stevenson	Jonathan Williams

and Mary Becker

The Brunswick Railroad Museum "Excavation":

Misty Bastian	James Castle
Cat Gasch	Linda Good
Mike Ricksecker	Kathy Rothenberger
Bert Richards	Herman (Mike) Stevenson

and Mary Becker

Introduction

All historical records are incomplete. There are still "remains" to be discovered and/or recovered throughout the histories of men and women. This means that we have a past reality that is still in the process of development. Reality is expandable. The data we currently have is a "sample". The question is: is it representative? The past is not "dead", but is it a repeatable past that can be experienced?

These questions (and their exploration) open the door for ghost research as a learning tool. Constructive fieldwork can increase our knowledge about the past, and its continuing presence today. This ability to produce new data and added meaning allows us to construct more sophisticated modes of reality. It enables us to re-interpret past life and forms of that past life today. Continued fieldwork, an "excavation" of its remains, rather than an "entertaining" pastime, gets us closer to the "truth" of what really lies "beyond the paranormal"!

Ghost research, however, cannot be the same type of science that physical science is. You can't mix "apples with oranges", or "people" (even "dead ones") with natural phenomenon and universal laws. A (not **The**) scientific method test theories through making predictions, and recording what results are obtained from those predictive models. But "ghosts" (like individuals before them) are not easily (or always) predictable (even under laboratory conditions). They certainly would not <u>increase</u> the level of predictability in death because they didn't (and don't) follow universal laws of behavior.

However, if we assume that a "ghostly presence" <u>is</u> unpredictable, and can only be perceived as a "paranormal event", then the cultural expressions that may be recorded in a haunting become relegated as incapable of being "scientifically" studied under controlled conditions. This makes ghost research irrelevant to serious contemporary scientific or academic inquiry. This gives the "apparitional experience" an insurmountable ethnocentric flavor, linking it to a materialist duality of reality ("normal" or "paranormal"). But what really is "normal"?

Following this line of reasoning leads us to inadequacies regarding the <u>possibilities</u> of a haunting, and the concept of a "ghost". A different perspective is revealed when we see an "apparitional experience" as a valid field measurement of a haunting, with references to particular cultural expressions from the past. The experience <u>can</u> become valid when recorded and observed by others and framed within a situation-defining socio-historical context.

If we haphazardly investigate a location as a "ghost hunt" (are "hunts" really organized-?), walking around (exchanging positions) and measuring the environment, taking random photos and doing periodic EVP, or just sitting around and viewing a monitor waiting for something (anything) to happen, there's <u>no</u> way of knowing that what was recorded, measured, or sensed is <u>representative</u> of "what" or "who" is actually there (if anything or anyone).It may be just a result of expectation, chance circumstance, or how one collects the data.

We must <u>learn</u> to increase the predictability of human unpredictability. We must learn to alter the unpredictability of an "apparitional experience". We do this through the use of context and learning about the "ins" and "outs" of different types of "investigative presence", and how each affects the predictability of this "apparitional experience".

Ghost research is much more than simply going to a haunted location and "learning the ropes" (or the workings of the "tech toys"): as a how-to guide to "ghost hunt"; how to scan, record, and measure; how to "debunk"; how to distinguish a "paranormal phenomenon" from a "natural" anomaly, etc. And it certainly is not a "training ground" (albeit a "haunted" one) for novice "ghost hunting" enthusiasts, or for children! Would a "real" investigator or scientist investigate with children, especially in many places deemed "unsafe" or "hazardous"?

If the past continues, then the form of that continuation can be viewed as more than a "ghost", and still remain a presence from the past. A "ghost" as something that remains is more than an "object" that is currently "unexplainable". It is also an "investigative presence" that is engaged in and with that "something" (or "someone") from the past.

It is the investigator making an appropriate ("contextual") gesture in the field, a voiced remark to "empty" space as social commentary that resonates with a particular past situation, the reading of an historical letter, the singing of a song, and the "witnessing" of what occurs during an enacted scenario. It is also positioning oneself into an appropriate historical stance, whether sitting at a table reading aloud that letter, or giving appropriate commands along a line of advance on a Civil War battlefield.

These behavioral acts and positioning also include "target objects": carrying a Civil War canteen as one "charges" along an avenue of approach where soldiers advanced toward a "key" position. It is doing a "roll-call" at a battlefield monument where the names of the dead are inscribed. It is playing the "rebel yell" while moving within concealed defensive positions along the former Confederate line.

All of these acts, logistical positions, and "target" materials are "strong" investigative presences that are used as exercises of engagement with "other" presences in haunted space. They connect present to past, collapsing "time" in the process. What remains, and "who" responds (and they do!) is a haunting reminder of how much more we can learn by not <u>disconnecting.</u>

This type of fieldwork is <u>not</u> re-enactment. It is contextual "investigative presence". It is <u>not</u> "theatrics". It's a form of ethnographic immersion, one that allows us to remain conscious of what is transpiring. This is <u>not </u> "ghostgasm", "wargasm", "techgasm", "grasping at straws", or putting out a "scarecrow" to help chase away critics, cynics, and skeptics.

"Presence", as act and professional stance, has a time-less "affect", especially when the "live" performance is reiterative. It "repeats" a <u>past</u> performance, invoking a "live" past presence. All documentation of this presence occurs at a <u>future</u> stage of the "afterlife" of the "ghost". It is deferred in time to a <u>future/contemporary</u> "investigative presence" that enables <u>past</u> presence to manifest. There is no termination or end of one time and the beginning of another. It makes sense, in investigating a haunting, to <u>forget</u> about time, by <u>remembering</u> an individual, their situation, and their <u>past</u> behavioral patterns. It is this past presence as a "real-time" experience that is the "afterlife" of a "ghost".

Legitimate research and relevant fieldwork as a social science perspective into the phenomena of hauntings, apparitional experience, and "ghosts" is faced with the same paradigmatic dilemma that anthropology encountered more than two decades ago. It is a question of fieldwork stance, encounters and interactions with the "other", and how to translate, understand, and transmit these encounters from the field, first to fellow investigators, and then to the general public. It is research that negatively implicates typical "ghost hunting" practices, the use of TV shows as a baseline for field methodology, and the framing of fieldwork experience as a "paranormal" event, all of which is perceived to lie outside the bounds of contemporary science and reality. The search for "truth" becomes a form of entertainment, rather than a learning process.

In traditional ethnographic fieldwork, the anthropologist is detached from locally-sensed experiences. The meaning of an act, event, and manifestation becomes a translation, not part of an immersive, participatory role as "insider" in the culture. This detachment from local reality means that the boundary between the fieldworker and the "other" (culture) is maintained. Similarly, in archaeological excavation, meaning comes from a translation of remains situated outside the contemporary (and excavating) culture. This maintains the boundary between the living and the dead.

Lost in both translations is the ethnographic commonality and archaeological continuity of human experience. We must go beyond the <u>detachment</u> from the "other". We have to go beyond the "observer", beyond the sparse physicality of what remains and what is. We must learn that the traces and vestiges , as the last perceived expressions of individuality, or a collective culture, and the haunted location (perceived as "ruin") is <u>not</u> the last (or only) expression of human presence and occupation.

Let's end the investigative quagmire expressed in our interactions with others we label as "ghosts". Let's end our detached positioning of difference (and indifference) regarding "ghostly matters". These contemporary biases are challenges to understanding what really occurs at haunted locations. It forces us to translate. It withholds us from establishing common ground (a shared experience) with those we consider merely as "phantoms".

If we continue to "hunt" (or "investigate") as most do, and if we merely perceive a haunted site as "ruin" rather than occupied, then the debris of remains as trace and vestige will <u>remain </u>mere "anomalies" of what once was (the "ghosts" of history). This means that we will never accept what we already have experienced as an expansion of <u>normal </u>(not "paranormal") reality. No sense of this normality will ever become established, unless we learn. Without learning, we will forever linger within the "box", the confines of something "paranormal". This is not to say that we do not acknowledge loss, decay, and death. We certainly do. But what we choose to focus on, and learn from, is our respectful intent to work with what and who remains from what once was. This calls for a different paradigm for ghost research!

What follows in these pages goes beyond the "paranormal". It entertains but it is surely <u>not</u> entertainment! My objective here is pure and simple. It is to show how, by questioning basic tenets of this "ghost hunting paradigm", we can (as anthropology has done) gone beyond the "traditional" reality of the field as genre, and the long established concepts of research, fieldwork, and interpretation, and arrive at an investigative position that is an "expanded normal", rather than a misrepresentation of what has long been regarded as "other". In the process of this exploration, we learn.

"Ghost hunting" has historically centered on a disassociation with the <u>commonality</u> of human experience. This commonality consists of both a horizontal perspective (the contemporary experience) and a vertical one (the historical record of ghostly presence). Given the importance of this vertical dimension, the role of archaeological field methodology, and the use of the "archaeological imagination" (Shanks 2012) become important aspects of the research and fieldwork. "Commonality" is centered on a difference (and distinction) between "life forms" (as physical entities) and "forms of life" (as manifestations of this commonality of experience).

This is an approach that goes beyond the "entrapment" of "paranormal" perception and thought, beyond the consideration of the possible "other" (apparition and ghost as "not quite human"), or "non-others" (demons, elementals, etc.). It is looking, thinking, and exploring past (at the past) with a consideration of <u>similarities</u> in human experience <u>and</u> everyday routines and activities. It is thinking of "forms of life", though <u>culturally</u> different from us, as being <u>included</u> in the definition of "us".

This research, and its fieldwork, is a particular orientation to methodology and its execution and the building of theories and their testing, that don't contradict the human experience of what acting human is like. It does <u>not</u> focus on measuring ambient deviations as part of <u>that</u> experience. It becomes a question of creating resonance between oneself (and ones memories) and the phenomenon one is attempting to understand in the field. It does not include, at least initially, us asking "them" questions (such as "Is anyone here"?) which do <u>not</u> resonate with the possible human experience of communication and interaction with past (human) presence!

Going beyond the paranormal becomes a re-orientation of purpose and execution that is a willingness to engage (by immersion) with another cultural world of human experience, following some form of commonality of experience (resonance). This is a way (quite effective) of producing manifestations (someone becoming present) of past human experience and memory. To take this "one step beyond" is to attend to intention, with intention fostering compassion and empathy. It is a moral and ethical approach to encounters with "presence" at perceived haunted locations.

Human experience, both present and past, become interlocking "fields" of memory practices. This is a form of "psychic unity", without the "paranormal" flavor (or the "parlor tricks"), and involves those aspects of common experience which makes us imaginable to one another. This empathetic resonance also fosters fieldwork identity as an "insider", not a socio-political dominant persona demanding action! It connects present and past feelings and thoughts. The connection, through resonance, provides context for that "apparitional experience".

Context is relative to the interpretation of this experience. Experience is set in context, connected by relevant relations to particular cultural acts and space. This creates a matrix of association. This book, and its re-focus on participation-immersion-empathy (P.I.E.) is about the "afterlife" of this past connection (context), and its associations to "apparitional experiences" at haunted locations. While a knowledge (and use) of context is important for fieldwork, it is critical that the contexts of past cultural behaviors (including habitual acts) form a baseline for P.I.E. field operational process: what contexts might (or might not) be appropriate for understanding the meaning of contemporary "apparitional experience". The typical "ghost hunting" acts and posture of "demand and command", so often seen and practiced in the field, are <u>not</u> contextually relevant (in most situations) to this experience.

During fieldwork, the codification of contemporary experience (the p.i.e. "acted out" as a cultural scenario and "in tune" with a particular cultural/sensual layer and historical horizon of haunting uncertainty) represents a means to test a hypothesis in the field under culturally-controlled situations. An appropriate context embraces an <u>expanded</u> situation beyond the contemporary, yet within the <u>contemporary</u> manifestations of <u>past</u> social situations and behaviors. The context allows for the <u>commonality</u> of present and past experience to emerge.

The demarcation of <u>discontinuous</u> worlds of conceptual differences as binary fragments (us vs. them; normal vs. paranormal; dead vs. alive; past vs. present) must be abandoned in fieldwork at haunted locations. The experiential continuity of being in the world at a haunted location, and locating forms of (cultural) life, means we must make appropriate contextual connections. This means that fieldwork beyond the paranormally- anomalous experience moves in the opposite direction: from these "ghost hunting" practices toward the recovery of continuity NOT difference. This challenges the concept of "something" (or "someone") manifesting as an "anomaly" to the contemporary actuality of a continuation of forms of life (as "afterlife conscious minds" in some instances).

We must question the materialism of the current scientific paradigm (cf. Sheldrake 2012) by expanding our perception and experience of the ordinary, habitual, and mundane. For it is this "ordinary", I propose, that still remains, and what manifests, at haunted locations. It's time that we learn how to interact in haunted space, not how to entertain those outside of it. This entertainment is easy, compared to learning what and who remains there at haunted sites. We must learn that it's only "paranormal" when fieldwork doesn't "unearth" a past interactive cultural presence that is human in character and not a "ghost"!

Anthropologist Clifford Geertz once famously stated that each of us is supposed to **"end in the end having lived only once" (1973:45)**. Life, in this perspective, is a movement toward history, something that was and is now past. Such a view eradicates possibilities by the perception that life always begins and ends at a certain point in time.

This is what the paradigm of materialist science tells us, and what the majority of academia teaches us. It is not, however, how contemporary reality is always experienced. Life forms may end or become extinct, but some "forms of life" continue. One of these "forms of life" is a sociable form, a cultural being. The possibility of this continuation creates an opening, not a "portal".

It is this "opening" that should be at the very heart of a ghost research that emphasizes the commonality of human expression and experience. This "opening" is a means to shift ghost research away from a fixation with "objects" (orbs, electronic devices) and images (the "ghost hunt" as it is mythologized on TV, on the internet, and in "fluff" ghost stories), towards a greater appreciation and understanding of the currents of human experience within which manifestations of "humanness" occur at haunted locations. It is this fluidity of "human commonality", rather than "otherness" (and manifesting as the "paranormal") that is the central concept of this book. Ghost research is the anthropology of a cultural "us", manifesting as layers of occupational presence at haunted locations.

We must go "beyond the paranormal". If one occupies a space considered haunted, it becomes a liminal position, between the present and some other time. To experience something beyond the present, we must move back to a past, and resonate with that past to understand its continuing manifestations. One must not maintain that present position by thinking, acting, and perceiving what is currently "in vogue". To <u>maintain</u> is to <u>remain</u> in that "liminal" ("paranormal") position. This does not gain us any movement of "advancing" the science of contemporary reality. For us, in ghost research, these manifestations cannot remain as "anomalies", something that is <u>not</u> understandable as past cultural acts.

Let the exploration and learning begin.....

Table of Contents

Photographs

"Ghost Story"

"There comes a time in every rightly constructed boy's life when he has a raging desire to go somewhere and dig for hidden treasure"

- **Mark Twain, <u>The Adventures of Tom Sawyer</u>**

As a boy, I sought that treasure in abandoned ruins of the anthracite coal industry in Northeast Pennsylvania (see Sabol 2007). For me, "play" there and then (the 1950's) became more than an innocent diversion. It was an enlightening journey into the inner "mines" of exploration, discovery, and recovery of the past (the past "Coal Rush" – now "dead"). It was indeed a "dig" for hidden treasures, both physical and imaginative, as I circulated within worlds that were now abandoned, many forgotten. There was no monetary value gained in these "diggings", but the experience was <u>"priceless"</u> as a learning tool. My education into ghost research began early, and it has haunted me ever since!

I grew-up and became an archaeologist, not a "treasure hunter". As I matured in my profession, I learned that not every "dig" required an excavation, and not every "ghost" occupied a "haunted house"! Today, I focus, at middle age, on "unearthing" more than archaeological <u>object </u>remains. I also recover subjects who have been lost, but who remain, performing acts of cultural habitual behaviors from memory and habit. I have come full circle, back to the "digs" of my youth, and the memories which continue to "ghost" my future actions.

As a contemporary archaeologist, I re-imagine the past as a potential future manifestation. Fieldwork becomes a "ghost excavation" into my past, as a memory of common experiences that "ghost" those of others, past. This "ghost excavation" is not a binary choice of revelation. It does not assign an "is" or "is not" haunted designation to a particular location. My fieldwork is a surface mapping of remains, both residual and interactive. These become **M.A.P.S.** defined as **"m**emories of **a**nthropological **p**resence in particular **s**ettings".

The M.A.P.S. is a guide for me in my explorations of "haunted" space. They help me to decide where I should "dig". They are the "ruined" spaces that I "excavate" in a "ghost excavation". There is one "mapping" that I remember quite fondly, though today I am physically far from its location. This is the Mahanoy Area Middle School where I studied, as well as my daughter, Melissa. And I lived not two streets away for 10 years. In these later years, both myself and "Lacy", my canine companion, walked the grounds around the school complex. The photograph on the cover of this book is the Mahanoy Area Middle

School, now abandoned as an educational facility, but not quite absent "past presence".

The building was once a stairway to a better life through education. Today, it symbolically serves as a "haunted" location in which to learn from ghost research. Though few climb those stairs today, the "ghosts" of former students and their memories saunter up and down those stairs, if only as residual presence.

But, inside the dark corridors, there is a legend of a "hunchback" and his dog. They are said to prowl the basement area of the building. Perhaps, the stairway should be used again, to instruct future generations about the importance of history, heritage, ethnic tradition, and education, to life in general, and to ghost research in particular. Is the story of the "hunchback" a symbol of that desperate need?

Mapping involves a particular concept of time at these haunted locations, especially institutions of learning. During a "ghost excavation", we set a baseline to experience a commonality of time, not sit and wait for it (time) to pass, or measure <u>duration</u> with scientific instruments that record deviant intervals. We "trigger" a particular manifesting situational <u>actuality</u> that is a specific sensual layer of uncertain situational time. This time is <u>not</u> a paranormal event. It is a time that is both anthropological and archaeological.

And it's time we discuss these disciplines, and their significance for ghost research. Otherwise, we will remain caught in the "time warp" that is a typical "ghost hunt". And that is a true haunting experience, full of "ghost stories" <u>without</u> contextual experience! And there is so much to learn beyond the typical "hunt" for ghosts! Just ask that "hunchback" if you meet him within the recesses of that abandoned educational institution!

Photo 1: A View of the Abandoned Mahanoy Area Middle School

The Background:

Why We Need to Change Old Thoughts into New Guises

Has "ghost hunting" changed the scope, shape, and significance of ghost research (as a legitimate learning and teaching discipline) since the exploration was first "voiced" by the "ghost hunter"? Has this "ghost hunting phraseology" changed through the years, the centuries? You decide…..

16th c. England:

Stay, illusion

If thou hast any sound or use of voice

Speak to me".

- **"Horatio", William Shakespeare (Hamlet)**

Early 21st c. America:

"Is anyone here with us tonight? Can you show us a sign of your presence!"

- **A "Ghost Hunter", Paranormal Reality TV**

No matter how you say it, it's still "ghost hunting"! Ghost hunting 101 began as a form of "entertaining" the masses, and it remains so today. For four centuries, ghosts have heard what should <u>not</u> be heard, failed to hear what they need to hear, and hear what most do not wish to hear! Is it any wonder that ghost research has <u>not</u> progressed in 400 years!!

"Attempts to raise the spirits of the dead have been made through the ages.....There are various methods of ghost raising.....Usually a circle is made with the aid of a compass.....The experimenter places a chair and table in the centre of the circle. On the table, he puts a candle, crucifix and Bible. He then sits himself on the chair.

.....He remains silently waiting the advent of the ghost, which may occur any time between midnight and two o'clock. The ghost may glide noiselessly into the room.....The experimenter, if he is not too overcome with fright, will speak to the ghost, and express his desire to help it in any way he can.

As a rule, however, the ghost makes no reply, vanishes, and leaves the experimenter no wiser than he was before" (Elliot O' Donnell 1969:16).

Elliott O'Donnell was the first true "ghost hunter", a man who devoted more than 50 years in the search for and documentation of ghosts, an investigative stance that began, not in the 21st c. (as **Most Haunted (2001)** or **Ghost Hunters (2004)**), but in the late 19th c. His more than 50 books not only made him the world's most readable ghost hunter, but the most read one as well.

Peter Underwood continued in the tradition of the highly popular and media successful British "ghost hunter", becoming England's most successful investigator as President and chief investigator of the Ghost Club of Great Britain. He subsequently formed the Ghost Club Society. He states that:

"A high percentage of ghost hunting time is spent simply watching and waiting and attending to routine matters concerning the investigation" (1983:94).

The typical "vigil" is this "watch and wait", framed by the deployment of various scientific devices to monitor the ambient environment. Here is his description of a "ghost hunt":

"I took with me that night various instruments for measuring.....and I attempted to 'control' a number of objects.....I sealed all doors and windows, ringed with chalk a number of moveable objects that had attracted attention in the past.....I even left pencils and pads of paper here and there.....Objects that had been

moved or disturbed in the past.....were all under special surveillance throughout the night. Powdered chalk was spread along the route taken by alleged apparitions.....and sensitive recording apparatus was placed where voices had been reported. Threads and wires were also strung.....at strategic places.....(1994:133).

Similar to O'Donnell, Underwood's field methodology is largely non-participatory, an inactive contemporary means of how not to (contextually) engage, establish co-equal status, and lay grounds for initiating and continuing communication with a past presence. Either they believe that "ghosts" are largely residual (non-interactive), or their methodology is too narrowly framed and executed. It does not involve "communication" or enable one to "entertain" a commonality of experience with the ghost(s) that haunts.

This "British tradition" continues today as evidenced by a recent book, **Ghost Hunting: A Survivor's Guide** by John Fraser (History Press, 2010). In this book, Fraser states that the "vigil" (though he does not call it this) is still considered the most common form of investigation. This "tradition" of "vigils" (with its "watch and wait" mentality, framed by instrument saturation), is also typical of many American "ghost hunts", largely influenced by what is observed on "paranormal reality" TV programs.

In **How to Research Haunted Locations (2012)** by Casper Waylin, he states the following:

"On-location, paranormal investigators sit quietly and wait. They set up their equipment conscientiously....Paranormal investigators manifest a great deal of patience as they go about their work...."

Answers to "ghostly" manifestations do not lie in waiting for something to happen, allowing the equipment to "work its magic". Listening in darkened, quiet rooms, I had learned long ago one hears sounds that are impossible to place, and one "sees" what the imagination conjures. The "vigil", as a "watch and wait", is a "dead" end! And the "hunt" for more sophisticated technological devices is not the answer to the legitimization of this research.

We've come a long way in "ghost hunting", a connection extending back to the 19th c., that has taken us <u>back</u> (not toward) results-based research at haunted locations. We have <u>learned</u> little about the "how" of investigating, but a lot about what we think we should <u>use</u> in the investigation. We remain today seeped in a "popular cultural haunting" focused on <u>popularity,</u> rather than the <u>expertise</u> of investigators. It's time we set the record straight, and move forward toward educating ourselves and others!

"Ghost Hunting" has for too long involved a particular "politics of presence" (Sabol 2009). Today, that "politics" has narrowed the field of inquiry by "ghosting" what most others do and say and, at the same time, what cannot be said and done. It also inhibits what can be made present and, consequently, what is kept absent and invisible. The "paranormal" TV shows and the "para-celebrities" (and their "entertaining" events) have further advanced a means of exploitation, rather than research. Unfortunately, with a strong "fan" base, they have held the power over the processes of the field of ghost research, such that this "power" and "politics" has become a critical issue for serious fieldwork.

Unfortunately, the old Victorian parlor tricks still haunt much of contemporary ghost research. Examples of this "haunting" are numerous, and include:

- Table-tilting;
- The use of a moving glass on a table to suggest contact through movement;
- Scrying; and the (currently in vogue)
- Flashlight test (to name a few).

Even contemporary technologies are used:

- The K-II meter as a "spirit answer machine";

- The ambient environmental measuring devices as indications of ghostly presence; and
- The "ghost box" (and its variants) as two-way paranormal "walkie-talkies".

David Rountree, in his book **Paranormal Technology:Understanding the Science of Ghost Hunting (2010),** has said of the K-II that **"it certainly is a lot of fun, but so far, I haven't heard from a single dead relative"** (2010:160). Regarding the Ovilus, as another device which claims to communicate with the dead, he says that **"it is of course, a hoax"**, and that it **"is a 'novelty item' only. In other words, expensive garbage"** (Ibid:161).

The "command center" at haunted locations, with its "watch and wait" mentality, and its techno-remote extensions to other spaces, has replaced the old Victorian séance room. But the contemporary implications are similar. Most attempts at communication do not emphasize context and resonance to particular times, cultural horizons, or individual personality and character profiles. Instead, "ghost hunters" rely on unproven devices to suggest presence, and "tricks" to invoke perceived manifesting presence. The result, most of the time, is "circus-like" entertainment. We continue to not learn!

Technology saves people from thinking critically about haunted space by re-arranging the experience.

The exploration of life becomes what Gil Germain (2009) has called the **"spiritization of humanity"** in which human qualities **"play a diminishing role in determining the conditions of life".** Ghost Hunters, "armed" with an "arsenal" of technological devices, become the "spirit beings" that transcend time and space through technology. They are the "ghosts" that haunt particular technological "haunted" space!

This overuse (and abuse) of technology re-shapes our social practices, moving it from research to entertainment, and has impacted how we "entertain" in the field of "ghost hunting", rather than learn about a haunting in ethnographic ghost research. We have <u>learned</u> this, time and again, in ghost research in the spaces of reality paranormal TV, in "ghost hunts", and in books by para-celebrities.

Haunted space has become, in these "ghost hunts", virtual reality which is equally ethereal because most "ghost hunters" merely play a simulated game of life (absence of something) and death (presence of nothing). Life does not necessarily consist of alternate realities. Technology makes it seem that way because it makes it (technology) productive as a new and alternative way of being in the world. But how relevant is this "alternative" when we are dealing with a haunting from a technologically-simple past?

This technological re-constructed reality is a measured "state of mind". It is not an external experience of a particular space and time. That space is not explored as performatively-authentic human cultural behavior. It is technologically-measured instead! Resonance is an important tool that allows us to "translate" the meaning of "apparitional experience" across time and space, not technology. With resonance, a manifesting presence is not "lost in translation". Resonance implies continuity, not separation. It is a different perspective, one that is technologically (by choice) "out of touch" with the present.

It's time to end the old forms of entertainment and begin a new mode (paradigm) of execution, a learning process to change the "face" and credibility of ghost research. It's still about making manifest ("unearthing") the past and past presence, and documenting this presence. But its focus must be centered on getting to know the past as it was – as it is. It's about a rationalized engagement with the commonality in this presence and the ordinary nature of the "apparitional experience". This is a mode of engagement that involves controlled fieldwork, the observation/recording of presence, and the media representation of the human experience and its commonality, both past and present. It is certainly not a "free-for-all" where anyone (everyone) can do what he/she wants! It is not about ego-driven "demands and commands". It is also not about "technophilia", a love of (and dependence on) the use of technology for "debunking" this presence, documenting absence, or merely recording "anomalies", without providing a baseline for understanding and meaning. In a recent book by John Fraser, **Ghost Hunting: A Survivors Guide (The History Press, 2010),** he points out this overuse of technology to monitor haunted space, suggesting that this overuse can induce fatigue, becoming counter-productive. He even states that it (the use of technology) is an obstacle for possibly engaging the ghostly presence.

And a "shape-shifter" is not a "ghost"! A "shape-shifter" is a being that can move between individual perceptions, assume identities, warrant different beliefs, and manifest during different investigative modes. At times, it can be an "orb", or a "rod". It can become a "mist", or even a "shadow". It can be a "dark figure" or a "light anomaly". These forms should <u>never </u>be the "ghost" of a former "living" person.

This "outdated" and "biased" focus on form is what much of contemporary "ghost hunting" is about. It must not <u>remain</u> what ghost research strives for, and can achieve. Though a perfect metaphor in this time of paranormal reality TV, we must not allow a "shape-shifting" entity to blur the true vision and purpose of ghost research as a learning process, and of what it means to survive physical death.

A move from an object of fear to a form of entertainment, or an emblem of celebrity status, must not be tolerated. We owe it to these "ghosts" to remember their human qualities, not their supposed characteristics. It is one thing to recognize and sympathize; quite another to exploit, taunt, and provoke.

A "ghost" is a redundant concept. It's been "us" all along! Why portray them as something else (a "phantom"), something more (a "supernatural" being), or something less (an "elemental")? The "ghost" in us is not a "paranormal event". It is paying attention to the learning experiences of an individual from our past!

There is no "ghost story" in these pages, or a "haunting" that continues as unanswered questions. This is no movie "serial",(a new "paranormal" adventure happening every week), or a prologue for Season two of a continuing paranormal reality TV show. It is the story of "us", those of us who remain from the past, and those of us today who engage, on a simple human scale, with those still present. This "excavating" the past is about re-animating it, as something and someone which has been lost and forgotten. It seeks, through mutual experiences and empathy, to make sense to people then, not now. This re-animation can be documented and subsequently told to others of us when we, as investigators, replicate past processes and human acts that resonate with particular occupational layers at haunted locations.

The implication is that cultural factors can become part of the physical make-up of the mind. If an "afterlife consciousness", as the mind of particular individuals, survives the death of the physical body and brain, and is interactive, then fieldwork performance practices must resonate with those cultural practices that were part of the enculturation process of particular individuals of a specific historical period. If a haunting is a past presence, perhaps containing "social and mental fields" (cf. Sheldrake 2012), then contemporary investigative practices resonating with these "pre-established" social/mental fields ("deep enculturation") would be "like attracting like". To interact with these fields, and the "afterlife consciousness" within them, we have to be like them, and immerse ourselves in their cultural world, not ours. This is the baseline that can change the current "ghost hunt" to a discipline of respectable social research and ethnographic fieldwork.

We cannot (will never adequately) make a connection between present and past if we fail to fully respect the human experience that continues of those who remain. A disrespect, breaching this continuity of humanness, comes about through an ideological, political, and technological imposition of present views/perceptions, attitudes, and our limiting notions of "paranormality". Haunted locations, as "ruins" of fragmented past presence, if left to this contemporary investigative paradigm, will remain a broken, lost, "paranormal past", one that is disinherited from us, and "alien" to the present.

To think otherwise, and remain fixed within the conventional box of contemporary "ghost hunt" practices is the real "ghost story" that continues to haunt those of us who are serious about ghost research as a legitimate academic field of inquiry, and those of us who remain from the past. Let us not lose the insight, or the vision, that ghost research can be (must be), if nothing else, admirable and redemptive fieldwork. As Rev. Louis R. Batzler, Ph.D. has stated:

"Ghosts can help to affirm the indestructible and worth of persons. Seeking to understand ghosts can provide insights and approaches to truth. Life visible needs life invisible for life to be indivisible and whole" (2011:69).

Let us envision (and perform) a way we can achieve this redemptive fieldwork!

Context: The Educational Key to Fieldwork, Investigative Performance And
The Documentation of Apparitional Experience

Anthropologist Clifford Geertz once said:

"Man is an animal suspended in webs of significance that he himself has spun. I take culture to be those webs, and the analysis of it not an experimental science in search of law, but an interpretative one in search of meaning" (1973:5).

The key issue in fieldwork in "perceived" haunted locations is not in identifying if a site is really haunted or not (a "reveal" or a "debunking" in "ghost hunt" mentality). What is important is identifying the context in which the site functioned as a social space, and which led to a haunting uncertainty. Then, using this context, we determine the "reality" of past presence at a site. Only in such "context" does the site become meaningful as a haunted location. For a haunted site must exist in <u>both</u> a time and social context.

There are no laws of social behavior, and as culture changes, from past to present, so do our interpretations of what remains. Our first obligation in research is to understand this meaning, and then, to the best of our ability, preserve the potentially informative context of these material forms of changed cultural presence.

In a haunting, this is a documentation of what was, and how much of what was has <u>not</u> changed (the "human" aspect), but also document the sensory manifestations that differ from contemporary cultural expressions of social life and belief. For example, regarding audio manifestations (beyond simple EVP responses), this would involve sonic elements that "lift-up over" the sounds of the contemporary landscape (as soundscape). These "soundings", as cultural expressions in a "web of significance", would be relative to a particular past time, place, and situation.

The greatest potentially-meaningful sources of a haunting are "remains" (as sensory traces) found in <u>undisturbed</u> contexts in which these remains, as past presence, remain embedded in the context in which they were <u>originally</u> associated. This means two things for contemporary fieldwork:

- Don't change the context – resonate with it;
- Don't investigate sites that are saturated by contemporary "ghost tourism" or have been popularized as locations of "para-celebrity" events or paranormal reality TV programming.

Context in ghost research is significant. It is not merely an association between a "feeling", and a recording (or photograph) indicating an "anomaly". The context in fieldwork goes far beyond that.....back (from the past) towards the future. Context must be established first, before fieldwork begins. It initiates with historical, ethnographic, and biographical research into the (often) multiple occupations of space at a haunted location.

Context is also an emergent property of knowledge acquisition, the determination of "what" (as vestige) and "who" (as trace) remains and continues to manifest at these haunted locations. The identification of context in the field form sets of relations that initiate two parallel processes:

- A present context, directed by us, as fieldworkers, using our own body of knowledge (historical, ethnographic, biographical) about the acts, behaviors, events, and usage of space of all occupations at these haunted locations; and
- A past context, manifesting from them, creating cultural situations and patterns, within their body of knowledge known at the time of physical death.

The intersection of these two parallel processes of knowledge come together, as accurately as possible (recorded and manifested) in fieldwork, during a "ghost excavation". This conjunction of <u>connecting</u> present to past <u>and</u> past to present (as resonating co-equal social and mental "fields") cannot be posed or enacted <u>without </u>the following contextual parameters:

- The setting: the social and spatial framework, including individual characteristics (age, sex, etc.) in which particular manifestations (as occurrences of "personhood") are documented;
- The behavioral environment: the investigative performance acts that initiate past presence communication;
- The language: the words, verbal tone, directionality of statement, and non-verbal cues that promote dialogue (I develop these into a storyboard of cultural scenarios); and
- The extra-situational context: the background knowledge for resonance.

Photo 2: The Burnside Bridge, Antietam Battlefield Storyboard

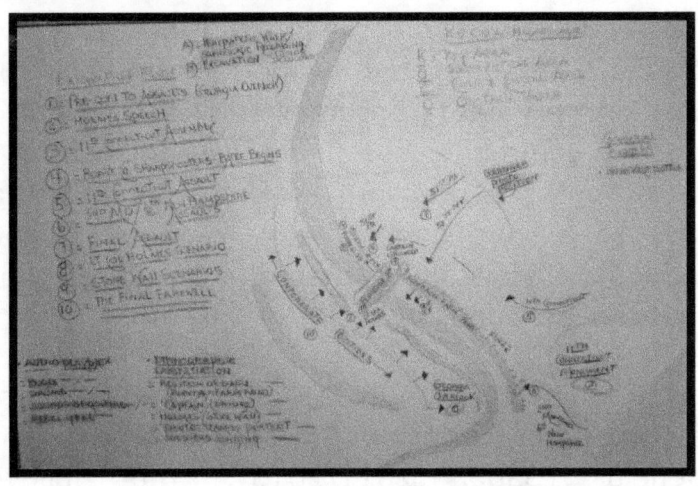

The context is by its very nature, present to past (and visa versa), allowing more possibilities for future manifestations as an acquisition of "what" and "who" remains from the past increases with continued investigative performances. This acquisition of data becomes an articulation of connective communicative fields, extending from the past, through the present, and into the future, <u>and</u> relative to a specific past actor (or actors) that is socially and historically- situated in particular spaces at haunted locations.

As an example of this, in the context of our "ghost excavations" at Burnside Bridge, on the Antietam Battlefield (Sabol 2011/2013), the following parameters were used:

- The physical setting was not the contemporary landscape but rather the physical terrain as perceived by the military commanders. This was (and still is) called the K.O.C.O.A. (Key areas, Observation areas, Cover and concealed areas, Obstacle areas, and Avenues of approach toward the Key area). In each of those 1862 militarily-defined spaces, we performed different scenarios that were contextual to <u>those</u> spaces. Within those K.O.C.O.A spaces,
- The behavioral frame was identified. This was the Inherent Military Probability (or I.M.P.), or how the soldiers would have acted (and what

they would have experienced) in particular situations within these K.O.C.O.A. spaces. We defined these as non-combative, non-lethal situations and experiences, since we could not enact the actual terror and cruelty of combat. Within this behavioral frame,

- A particular language (including non-verbal) was used. Examples of this were "roll-call" in a cover and concealed area, and a search and rescue, enacted by female investigators (as "nurses"), in the obstacle areas. All of these contexts were based on

- An extra-situational frame, defined by the mid-19th c. belief in the "good death" . An example of this was exemplified by the search for missing relatives/loved ones by female family members during and after the war. At Burnside Bridge, female investigators, portraying female relatives, searched the area for the remains of Lt. Colonel William Holmes, 2nd Georgia, who is still buried near the bridge in an unmarked grave.

Our "ghost excavations" at Burnside Bridge are documented, and the reader may want to read about and hear some of the audio we recorded there during our "excavations" at www.ghostexcavation.com. As we continue to "excavate" the "hauntscape" at Burnside Bridge, we are learning more and more about how the hauntings are manifesting there. This learning process is explained in a later section of this book.

Photo 3: Burnside Bridge, Antietam Battlefield (Maryland)

Photo 4 : Context and Performance at Burnside Bridge

Another use of context is our "ghost excavations" at the Knickerbocker Hotel in Linesville, Pennsylvania, a former hotel containing multiple occupational layers of haunting uncertainties. Our learning "experiences" at the "Knick" are also summarized in a later chapter in this book.

Photo 5: The Knickerbocker Hotel

A further example of the use of context is our "ghost excavations" at the Brunswick Railroad Museum in Brunswick, Maryland, located approximately 15 miles east of Frederick, Maryland. The results of this "ghost excavation" are also summarized in a later chapter of this book.

Photo 6: The Brunswick Railroad Museum, Brunswick, Maryland.

During each of these "ghost excavations", the contextual parameters were used to develop a storyboard, within which various cultural scenarios were identified and described, and the results noted, one at a time (whether they were successful or not). These results formed a baseline for future iterations of similar and expanded scenarios at the site. When a particular scenario "unearthed" what was believed to be an emerging manifesting presence, in any sensory modality (visual, auditory, tactile, olfactory), the "communication" is immediately followed-up and continued until communicational responses (as contextual "ethnographies of communication" or E.O.C.) ceased. During this process, we used RT-EVP audio recorders to determine presence, and initiate immediate communication.

In our "ghost excavations", the storyboard is not completely executed due to the amount of perceived manifestations that may occur, which required some form of immediate contextual response from the field team. These immediate responses helped to maintain the integrity of both context and process (as communication flow) during the "excavation", thus enabling future iterations of an ever expanding storyboard on subsequent investigations.

A "demand and command" approach ("Do this......Can you move this.....Turn the flashlight on....." etc.) as provocation makes for great television entertainment, helping to alleviate the boredom of repetitive acts from one paranormal TV show to another. But the activity is an ineffective tool for serious research because it creates an un-controlled, non-contextual situation. Besides, it is unethical as a cultural mechanism (except in a few select instances) for interaction and communication with (former) social beings.

A good example of a "manifesting presence" within a contextual frame is a transductive E.O.C. This is an EVP recording (with simultaneous manifestations of other sensory modalities) that was unanticipated but relative to a cultural scenario that was enacted just prior to the transductive manifestation. At Burnside Bridge on the Antietam battlefield, we recorded a transductive E.O.C. from a "soldier" that clearly asked me directly the following: "Captain, is that you Captain"?. This EVP can be heard on our website at www.ghostexcavation.com. It occurred, along with some visuals of "shadow" movement, as I was crossing the bridge, reviewing the storyboard, walking toward a group of investigators situated in the middle of the bridge. This transductive manifestation was contextual because it manifested shortly after I did a "roll-call" of the 11th Connecticut, who enacted the 1st assault toward the bridge. I propose that this EVP was questioning if I was the "Captain" (as the officer in charge) who did the previous act, the "roll-call". An alternative explanation to this recording might involve a communication between two "ghost" soldiers, and not directed at me.

Context is extremely important. It helps us to substantiate the "matrix" of associations of a manifestation, the relation between what we do (as fieldwork) and what (and "who") becomes a result of our field performance. How confident are we that these manifestations are responses from the dead? The answer is framed by the context and matrix: it occurs within the communicative flow of the scenario we just enacted, and immediately recorded on our RT-EVP audio recorders, and co-verified by other audio equipment used at the site.

Biologist Rupert Sheldrake, in a pioneering study (1999) focused on the extra-sensory ability of pets which anticipated when their owners would arrive home. I myself have experienced this phenomenon with my cat (Tazia) and my dog (Lacy) who both waited at the door moments before I arrived home from work, even when I was early or late. Sheldrake concludes that there is an empathetic bond between the owner and his/her pet that enables interactive telepathic communication. I propose that this "empathetic link" can be useful in ghost research. In a "ghost excavation", we do a similar empathetic link through our contextual scenarios with specific historical individuals in particular spaces.

Context is also important for determining the scope of fieldwork in ghost research. It is to this them that I now discuss.

Learning About "Presence"

At a haunted location, in that spectral space, there are various levels and intensities of "presence", <u>all</u> of which are important for understanding the meaning of ghost research. And I am <u>not</u> talking about a "ghostly" one! I am referring to "presence" as <u>investigative</u> "presence". These levels of "presence" go far beyond a "ghost hunting" appearance, and have little to do with "presence" as seen on paranormal reality TV programming.

The first type of "presence" is a "weak presence". This conforms to Fischer-Lichte's (2008) concept of simple physical presence. It is defined as merely "being there". There is little doubt that this refers to many "ghost hunting" presences at haunted locations. It is characterized by "walks" around the site, measuring the ambient nature of the space and its deviations, and asking such absurd questions such as "Is anyone here with us tonight?" This <u>is</u> a form of profanity that is noted for its lack of cultural and historical context, its "inhuman" connotation, and its inability to resonate with past space and situations. And it is a waste of investigative time because it "identifies" the investigator (s) as an "outsider" to "who" may remain from that past! Without context (doing something that resonates with situations in past space), the question arises as to "what" and "who" may be present.

A second type of "investigative presence" is "strong presence". This is Fischer-Lichte's concept of the **"ability of commanding space and holding attention" (2008:96)**. This is a "presence" that is empathetic to the emotions, mental states, though processes, and personality characteristics of the "ghosts" who may haunt a particular space. As this "presence", the investigator would not be perceived as being an "outsider", but rather a dramatic figure "who" becomes part of that space, a particular situation, and that time in the "life" of the "ghost". This is the type of "presence" that we try to emulate during a "ghost excavation". This type of "presence" differs considerably from "ghost hunting" field acts of scan, measure, and monitor. A "strong" investigative presence is goal-oriented, intentional activity with the use of a particular "model" in that intervention. For example, on a Civil War battlefield, the "model" that we use is "Inherent Military Probability" (I.M.P.) of the "culture of war", or how the soldier would have reacted (and what he would have experienced) in a particular battlefield situation and space.

This "investigative presence" would "bridge" the time between present and past because as being "present" (in that past modality through resonance) one would connect with an established field of (past) activity. And it would be recognizable to the "ghost" as compatible to a past memory pattern. In this way, the investigator learns during fieldwork what is

<u>appropriate</u> behavior (and what is not), as the "ghost" learns to <u>react</u> to that appropriate behavior.

The investigator becomes a "strong presence" because of this ability to re-occupy and command a particular past space, and to "attract" the attention of the "ghost" by this <u>appropriate</u> behavior. The "ghost" would sense, I propose, a certain power and frequency that emanates from the investigator, possibly perceiving it as a source of energy (without feeling overwhelmed, afraid, or other negative reactions). To the "ghost", another "presence" is confirmed, and occurs as a "strong" experience of <u>common</u> "presentness".

This "presentness" is <u>not</u> based on "passivity" (a "weak presence" of doing little or nothing relative to what occurred in that space), or something that is <u>beyond</u> the "ghost's" recognition (such as using modern tech devices). A "strong presence" is relative to the performance (resonant and contextual) of the investigator (not the tech device). Performance becomes the generation of a "targeted" energy flow that is <u>transferred</u> into haunted space. This transference of energy is meant to <u>direct</u> the investigative process toward a particular "ghostly presence", at a specific layer of uncertainty in haunted space.

This quality of energy flow through space was theorized quite extensively by Eugenio Barba (1986) in his discussions on artistic articulation. He states that these practices (transmitting "presence") are developed and actualized by performers in many diverse cultures to direct the energy within themselves, allowing it to circulate through space and connect to other presences which are also generating their own energy. This concept of "embodiment" (transfer of experience as energy) has been used as part of the "archaeological imagination" (Shanks 2012), an appropriate investigative tome for ghost research (as the recovery and documentation of past presence).

Finally, there is a "radical" concept of "investigative presence". This is an exaggerated form of "embodiment". A extreme version of "ghost hunting". Though it may use the context of a situation in past space, it does not necessarily involve an <u>empathetic</u> performance. This performance does not <u>usually</u> transfer well into haunted space or the "ghosts" that may haunt that space.

Sometimes, the performance is meant to taunt. At other times, it is a "demand and command". In "extreme" cases, it provokes <u>unnecessarily</u> to get a response. This may work in haunted prison space (to an extent), but it is <u>not</u> appropriate (or humanistic) for <u>most</u> haunted space. What I have learned about "investigative presence", in more than four decades of fieldwork, is that a "strong" presence is most appropriate in the majority of haunted locations.

It must be emphasized, however, that this "investigative presence" is not personal charisma, and it does not simply emerge because the investigator repeats, no matter how "accurately" and precise, a past action, event, or "mood". It becomes present through a conscious awareness and perception of the purpose of the scenario in the moment of its enactment in fieldwork.

"Investigative presence" emerges, and becomes significant, if and when there is "unearthed" a manifesting response relative to that performance. It is when "investigative presence" begets an "apparitional experience". For me, as an archaeologist, this occurs during an "excavation" of haunted space. The "excavation" is an attempt to get back to where actual past presence existed as contemporary presence. It was when the "ghost" was still wholly human. In contemporary reality, a "ghost", I propose, is what's humanly left of the presence of the past.

Ghost research should be a re-animation of this past presence, as it occurs in a haunting. It should not be a form of entertainment. Fieldwork, and a "strong" investigative presence, is a means to bring past presence back into "life", providing a future direction and learning experience for investigators. This process of "re-animation" becomes the "second life" of a haunting through mutual presence. The "strong" presence of the investigator, through performance practices, allows the presence of a manifestation to become a symptomatic trace of a past situation or event. The result of this process can become a science of past material traces, and not a "paranormal ghost hunt"!

What we should be doing in ghost research is working on a "tool-kit" that deals with these traces of the past ("presences") as a kind of genealogy: how "ghosts" have come down to the present and manifest. This is in contrast to the "traditional" sense of "ghost hunting" as principally the documentation of changes in the contemporary environment. How important is this measurement when the past already exists as a field of presence?

These measurements neglect the engagement of presence to someone becoming present. This is not just the experience of it, but also the forensic work of learning from the engagement, and lessons about the "what", "who", and "when" of a haunting. That is learning about "presencing", the fieldwork as a valuable tool about the existence of "presence".

The Educational Scope of Ghost Research as a Learning Tool

"Ghosts, with their tendency to unsettle our comfort zones, mitigate against complacency, and encourage us to keep asking questions"

- **Catherine Spooner, Senior Lecturer in English Literature**

To merely observe (including scanning the environment with electronic equipment searching for "anomalies"), I propose, while waiting for something to happen without participating in acts of behavior "ghosting" what once occurred at haunted locations, is to take the "life" out of "who" still continues to "perform" acts of cultural behavior at haunted locations. In contrast to a "watch and wait", performance-based fieldwork is a <u>controlled</u> study of the "afterlife" of former "life forms" occurring as sensory materializations of "forms of life" from a <u>particular</u> historical period and a specific cultural and/or ethnic horizon.

This <u>is</u> a discipline that becomes the study of past cultural experience (as historical ethnography) that unearths a "deep enculturation" experience as a layer of haunting uncertainty (an "archaeological imagination" (Shanks 2012)). If <u>any</u> of the manifestations that occur at haunted locations are <u>true</u> past presences of the minds of "dead" people, then, as David Fontana has said, **"many of the most fundamental laws of science would have to be rewritten" (2009).** This "rewriting", according to Fontana,

"challenges the supremacy of material science, reducing it from the ultimate authority on life, death, and everything to simply the science of material things" (quoted in Noory and Guiley 2011: 274).

For me, this changes the "science" of ghost research to an anthropological/archaeological discipline that emphasizes a "commonality of experience" in field performance acts, ones that resonate with past behaviors of specific individuals in particular spaces. Fieldwork becomes a "tuning-in" to a pre-established (i.e. already existing **NOT** "paranormal") past cultural field of social acts, behaviors, and mind sets (cf. Sheldrake 2012) that may remain (however fragmented) at haunted locations.

These past cultural acts are a "field" of resonating potential of what remains ("in ruin") after events, situations, and individuals have usually passed (and become "past") and have been assigned to the historical record. Those remains are past social fields (behavioral/cultural acts) and mental fields (states of "afterlife consciousness") in archaeological context (layers of surface uncertainties) that remain. A "ghost excavation" works <u>with</u> what (and who) remains of (in) these ruined fields of past remains.

We must attempt to move beyond the "hunting" mentality that is so common in ghost research. It focuses on an intimate engagement, through archaeological field practices and the "archaeological imagination" (cf. Shanks 2012), "digging" into the basic principles of performance, immersion, and empathy (P.I.E.). In the "dig", there is less focus and concern with how we physically "dig", and more with how we interact when we "dig". During a "ghost excavation", we can have our "p.i.e." and learn at the same time.

In fieldwork, we do not have to make interpretations at the trowel's edge (Hodder 1997), or in the physical negotiations that take place around it. We must not flatten the haunting to a single historically-placed phenomenon, unanchored by socio-cultural parameters. We also cannot think of a haunted place as a site that belongs to the past, as opposed to multiple spaces of activity that are varying tiers of fragments and traces that remain on the surface of haunted space.

The problem in "ghost hunting", one that still has not be learned since the time of Elliott O'Donnell, is that <u>subjective</u> temporalities (a haunting "presence") cannot be reduced to <u>objectified</u> measurements (a drop in temperature; high EMF; an "orb", etc.). This continues the binary of subject/object, so typical of materialist science, without considering the "ghost" as an historical agent (and cultural being) with <u>purposeful </u>behaviors.

This means that how a site is "excavated" is <u>not</u> important. It is about giving our "excavations" a resonating translation and transformation from present to past. This means that new strategies of recording in context must be developed as we learn to interact with the past. What is troubling is the perception of many "ghost hunters" that this "translation" and interaction at haunted locations is <u>not </u>a problem. It is! This is aimed at those who think they are operating "outside the box" of traditional "hunts", and any problems are "fixed" by the application of more scientific devices and their correct usage. This simply reduces the role of fieldwork in ghost research to mere data collection from these electronic devices, waiting for (instead of interacting with) the manifestation, which are commonly identified as "anomalies" in the subsequent "reveal".

The exploration of haunted space, as performance-based fieldwork, is, I propose, a more embodied, humanistic, socially-attuned, participatory approach to research on the "afterlife" at haunted locations. This fieldwork, by uniting present and past socio-cultural and mental fields of patterned acts, link visible and invisible worlds of contemporary reality (and their associated occupations) into <u>one</u> field of mutual experiences. Such a field methodology recognizes the important communicative position of the fieldworker as an active, contextually-sound participant in the process of acquiring knowledge ("becoming educated") about the continuation of past presence at a location.

This methodology is used as a <u>human</u> "tool" to document the presence of the past and the "afterlife consciousness" of individuals who may linger at haunted locations. This focus on a "commonality of experience" and behavioral patterning from the present to the past (and visa versa), gives relevance, voice, and meaning toward future manifestations of a <u>similar</u> nature at haunted sites. In the process, <u>we learn</u> what it means to document haunted space!

A Lesson From Cultural Anthropology

Anthropology is a science where observation, description, and <u>purpose</u> are entertained, not as forms of entertainment. The purpose of anthropology, I believe, is to observe and describe the conditions and potentials of human life through cultural expressions of belief as they manifest outward in the world.

Ghost research, as a form of ethnographic work at a haunted location, is a participation in the lives of those one is studying, even though these individuals may be physically "dead". This "participatory" approach to "culture" (and a "ghost culture" I propose) has a long-standing anthropological tradition, beginning in the early decades of the 20th century.

Equally historical is the use of one's own experiences in the gathering of data in the field, and as an interpretive tool. Anthropologist Tim Ingold (2011) has argued that to merely observe and not participate is to take the life out of ethnographic fieldwork. This is an apt lesson to be learned for those of us who investigate haunted locations, and the traces and vestiges that remain of past cultural expressions ("ghost culture").

Merely measuring and noting ambient changes, watching a video monitor, or acting out of context is to <u>maintain</u> the past as something "dead and gone". Participation <u>in</u> context is an integral part of understanding the meaning of a manifesting presence as "apparitional experience" at haunted locations. Experiences that do <u>not</u> fit within the frame of what occurred in the past in a particular space should be <u>excluded</u> from ethnographic analysis of a haunting.

These conditions and potentials, as "human cultural life", do not begin here and end there. Some forms of (this) life keep on going with space, situation, and purpose clearly in mind. An individual (not a "ghost") is the producer of this process which creates history as biography. The experience is humanly possible, but the biographic history is uniquely human. It is this unique biographic human cultural history, as space, situation, and purpose that we explore in a haunting.

As a trained anthropologist, I am always between something and somewhere, even when I am physically stationary. I was enculturated into "Western Culture", but I study other cultures. I live in the USA, but I have traveled (as that anthropologist) outside those geographical boundaries. I am situated in one space, but I observe and record the manifestations of other spaces. I "inhabit" multiple liminal positions, a <u>perfect</u> "location" for researching a haunting.

This concept of position begets a philosophy that combines two or more apprehensions of reality. A "ghost" co-occupies a similar circumstance, situated between two "worlds", two spaces, and two (perhaps multiple) temporalities. And a haunted site is a location of ambiguity, something between absence and presence, silence and sound, and that moment when something unexpected just happens. The "liminal", once thought a "primitive" rite of passage, is a normal position for all of us, and that position illuminates the experiential multiplicities of human social life as actuality today.

People (and investigators) and the "ghosts" of place in these liminal positions are actors that must find paths to walk toward or away from. There are individual solutions that must be met – both in space (walk to a different space) and in time ("dig-down" or remain bound). What steps are eventually taken becomes a "badge" of identity.

In liminality, the researcher is no longer an objective observer that records and measures. Instead, participation is required. And the "ghost" is not an actor to an empty theatre, or in the audience merely viewing what is already known. Fieldwork is responsive (and responsible) actions to behavioral acts. These become "re-acts", a flow away from "center stage". It is to go beyond the known edge of what is the contemporary situation.

The field of cultural anthropology, reinforced by theory and its history, involves a journey (until relatively recently) into unique, non-Western and distant cultures. Its fieldwork thrust was an attempt to understand the social realm of groups, and their cultural beliefs that went beyond the boundaries of a particular time, space, and individual life, and of what was currently known. This is the adoption of a strategy of "assuming continuity" involving ethnographers in the field who **"emphasize the inevitable experiential continuity of the human world irrespective of time and place" (Palsson 1994:12)**.

The ethnographic/historic commitment to the notion of discontinuous worlds (the "West" vs. the "rest"), however, had produced multiple problems of cultural ambiguities for the fieldworker: how to "translate" these other cultural beliefs and their social realms toward an understanding in Western terms (for both the scholar and the general public), at the expense of those who were studied. Ethnographic fieldwork became a presentation of culture as text, not learned experience. Anthropology became a reading of this text which was usually not understood by the subjects of the study.

This "reading" of a "translated" cultural "text" assumes that the observer (the anthropologist as ethnographer) is removed from the society under study, and, therefore, the phenomenon that he or she is attempting to study. The result becomes a "hunt" to bring order and meaning to what is being observed, from a contemporary "Western" point of view. The end result is a "translation", not what may be occurring in the society from their point of view. If the ethnographer is merely "translating" for a Western (and contemporary) audience, what criteria then constitutes an authentic and successful translation?

Because of these problems with "translation", there evolved an alternative approach in fieldwork methodology. Some called it a **communion of experience" (Ingold 1994)**, one that went beyond the role of the anthropologist as merely an observer and recorder of data in the field. Others labeled it the power of **"resonance" (Wikan 1994)**. Both of these terms address aspects of an ethnographic model of empathic communication, which attempted to eliminate the concept of a "detached observer" in fieldwork who "translates" the field data for others to understand. This problem of anthropological "translation" is an important lesson we can learn from in ghost research.

This involves the use, in most "ghost hunting", of "translating" the measurements, readings, and recordings by electronic devices as indicators of a "haunting" and "ghostly presence". It involves the "mis-translation" of "things" as "objects" (orbs, shadows, drop in temperature, elevated EMF, etc.), rather than "manifestations" as indicators (in context) of past cultural behaviors and <u>still</u> human ("subject") presence. Since when are "ghosts" these "objects", rather than human "subjects"? This replacing of a "subject" with an "object" is a <u>contemporary</u> "translation" of haunted space.

Ethnographic work today seeks to use various elements of social situations that may be encountered in the field, looking at these situations from a perceived "native's point of view". Implicit in such an approach is that the data obtained might produce a reliable "map" of the culture under study. Should ghost research at haunted locations follow this cultural template? To learn about a haunting, from the "ghosts point of view", I propose that it must! This is an important lesson that we can learn from the lessons of inaccurate ethnographic "translations".

I have used various contexts for this cultural "mapping" at haunted locations during fieldwork for the past several decades. Some of these include:

- The K.O.C.O.A. as an imagined "hauntscape" on Civil War battlefields (cf. Sabol 2008; Sabol 2011: Sabol 2013);
- The use of identifying "soundmarks" (bells; attendance/roll-call; contextual lesson plans, etc.) at haunted schools;

- The use of "staged" performances at haunted theatres;
- The use of "hospitality" as a cornerstone to "excavating" a haunted hotel; and
- The use of incarceration philosophies and corresponding architectural modifications at haunted prisons.

In all of these imagined cultural "mappings", there is a commonality of experiences, a resonance that allows a link between present to past to be established. It allows present reality to expand back to include the past. We learn to be a part of that past, derived from what we learned as part of the present. We co-inhabit two worlds, surely a liminal position.

These "mappings" form the contexts through which relevant relationships and social frames can be used, such that the "subjects" (as "afterlife conscious minds") can act out (once again) part of their lives as they experienced them in the past. The documentation of these manifestations becomes **"an anthropological voyage"**, as anthropologist Hortense Powdermaker once remarked, of ethnographic fieldwork that occur within various hauntingly-uncertain layers at a haunted location. The fieldwork offers ethnographically-inspired situations and associations, and the frame of culture, as an investigative means in which an investigator can mobilize questions and discover techniques for arriving at answers to the haunting uncertainties that may be manifesting at a given location.

Central to this "ethnographic mapping" is the concept of "transduction", as the transfer of communicative "signals" from one venue (investigative) to another (the haunting). The transductive act initiates what anthropologist Clifford Geertz has called **"webs of significance"**. This is a layered, multi-related network (or matrix) of meaning that is carried (from present to past) by specific historical dialogues and words, acts, conceptions of space, soundmarks, and other resonating elements. This becomes an alternative and <u>learned anthropological</u> vision of becoming present in haunted space. The "ghost" becomes a cultural being once again!

Death and the "Lives" of a Haunted Location

All haunted locations don't have much of a "life" before they were "discovered" in their present context, as perceived to be inhabited by ghosts. All our knowledge, exploring forward, whether measured, recorded, or speculative about past presence are outcomes of this initial discovery. This presence, however, may mean different things to different people, be they "ghost hunters", ghost tourism businesses, reality TV producers, scientists, skeptics, or the general public. Thus, haunted space can have parallel lives in various present contexts, depending upon the varying movements (and experiences) involved in the discovery.

To study such present stories of "live" encounters of "life" and their contexts requires an ethnographic approach: an immersion into the social beliefs and acts of each of these discoverers. It also means analyzing how these discoveries can affect both the ghosts of place, and the impact on haunted space for future explorations. In the present, an observer records the space, but in the future, this observer may become the observed. Our acts leave traces of us as residual elements on the environment, more so if we do not resonate with what already has occurred in the past.

Whatever our motives or belief systems are, we must learn to re-learn the past, such that we can become an "insider" in haunted space. We must become a "recoverer" of what is now, and not a "discoverer" (as "outsider") to "who" remains. We explore to recover past ground, not discover new lands. It's time we re-learn old things, such as:

- Find out more about our way of life as we are doing (and also doing wrong) when we explore the past at haunted locations;
- How we can transform "forms of life" into forms of documentation that find agreement among different contemporary "life forms";
- How can we learn what is past cultural life that is natural, distinguishing it from present natural life; and
- How do we establish past routines and habits that are <u>recognizable</u> by "ghosts", and which may stimulate recall which <u>we</u> recognize, as a manifesting from of life.

To do ghost research, explore haunted locations, and record manifestations of past forms of (cultural) life, we must learn to recognize certain things as representing a haunting by a human "afterlife consciousness". The bottom (base) line, after "digging deep" into the past, is that the "lives" of ghosts at a haunted location in the present is not nearly as exciting (or complete) as those lives were in the past. We must learn to accept and be content with "forms of life" that are partial, fragmented, and decaying. A haunting is a "life" in <u>ruin.</u>

But, those past lives, as products of particular cultural worlds, are the <u>direct</u> (not "paranormal") outcome of <u>contemporary</u> "forms of life" at these sites discovered to be haunted. Only with a secure identity as to their past origin can ghost research implicate a manifestation as a plausible connection to individuals of a particular past and situational context. This is to recognize these forms of life for <u>who</u> they <u>are</u> and therefore <u>who</u> they <u>were.</u>

Learning is a challenge, especially if our "teachers" are (the) past. If we claim to know something previously undocumented from the past, rather than simply believe it, think it, or wish it, we must be willing to confront challenges and alternative views. This is not being polite, a courtesy, a form of scientific openness, or even "paranormal unity". It must be a prerequisite to distinguishing between that acquired knowledge and opinion. In the cases that follow, the knowledge that was acquired in fieldwork at haunted locations was a learning process on how to interact, through trial and error, with cultural beings from the past.

Finally, this investigative process was the socialization of an archaeologist to past subjects, not objects. What remains of the past at haunted locations are <u>not</u> orbs, EVP clips, or a measured ambient deviation. They are cultural expressions of forms of life still not past. Fieldwork at haunted locations is not merely the sum of applied method and theory to space. It is also an experience of lives that are still significant in ways other than what many purport it to be as "paranormal events", "anomalies" or "empty" space.

How Fieldwork Comes To Be a Learning Experience

A "live" encounter with a <u>real</u> "ghost", as an authentic "apparitional experience", is a <u>mutual</u> learning experience:

- The fieldworker learns how to produce a manifesting presence through resonating cultural acts; and
- The "ghost" learns through recognition (identifying the investigating act as culturally sound) and recall (remembering a past act) what still survives in conscious memory.

Those who present past presence (as "ghosts") to others must <u>also</u> learn their obligation and responsibility to represent the most likely reality of this encounter. This represented encounter must <u>never</u> be a conscious manipulation of data, or a trace summary of historical inaccuracies that promote contemporary causes, such as entertainment ("ghost tourism") or self-marketing (for paranormal reality TV prospects). These "re-constructions" of reality may involve subjective perceptions or personal scientifically-unfounded beliefs, framed by biased confirmations, such as the <u>presumed</u> authenticity of most "ghost tech" equipment.

The "re-construction" of most "ghost hunts" is not how the past was constructed, or how individuals really lived in the past. There is, however, not <u>one</u> past that is known to <u>all.</u> There are as many interpretations of past events and situations, as there are sciences to interpret it (cf. Sheldrake 2012). As one form of "interpreters" we, in ghost research, must base our work on the most reliable up-to-date information and data available. This requires reading and <u>learning</u> the literature.

This is an <u>obligation,</u> and <u>not </u>a choice! To think and act otherwise would intentionally create the image of the past (and ghost research) that has little credibility to what may be true, and fails to provide meaning to what (and who) is manifesting at haunted locations. It is, to quote anthropologist Carol Delaney's phrase, **"to bear the burdens of one's observations" (1988:292).**

There are times, largely unspoken in "ghost hunts", when silence, rather than an EVP, is meaningful. The problem is that "ghost hunters", in large majorities, tend to "over-communicate" their <u>own</u> presence in the field. They do this by assuming the stance of a "ghost hunter", rather than taking a resonant position in the past society under study. This is reflected in their dress, mannerisms, speech patterns, and technology.

We must learn to re-learn:

Fieldwork in ghost research means we must be "experiencers", rather than receivers of measurements. We must be "creators" of meaning, rather than translators of anomalies. We must become "actors", rather than distant observers of video monitors. We must have intention to get attention. We must be available, and not allow the instruments to take our place in haunted space. And we must pay attention to what we practice, not what the instruments register.

Our identity must be constructed, not formed from the equipment in use. Fieldwork is not a containment, a box, filled with devices. It is an "opening" that we "excavate" through acts of resonance. It is about being there in the past, not remaining here in the present. Learning is the understanding of the consequences of our actions, rather than searching for the causes of a haunting. What began as a "hunt" continues as a recording of "forms of life". What once consisted of fieldwork as a measurement of space becomes a documentation of cultural behavior.

In this book, I will show, through actual case studies, how we learned about past presence (both its presence and absence) at haunted locations. This was achieved through repeated iterations of investigative fieldwork at these locations. Fieldwork in ghost research is an ongoing learning process into understanding the contextual meaning of a haunting.

In general, however, there is increasing criticism with the use of dichotomies to understand the haunting past. These impose a contemporary way of looking at the past. This is a view of subjects (as investigators) and objects (as manifestations). And in "ghost hunting", there is mainly a confusion that exists between who is the "subject" and what is the "object".

A haunted site is measured, recorded, described, and photographed as an "object" of study, and interpreted by the field team (as "subjects"), but is the "subject" <u>clearly</u> divisible from the "object" of analysis (the "ghosts")? "Object" physicality is "exhibited" as "light anomalies", EVP, subjective perceptions, and notations that indicate the state of the ambient environment which may or may not indicate a "ghostly presence" (drop in temperature, anomalous change in EMF, etc).

Since when is a "ghost", as a past human presence (a "subject") become an "object"? And don't say it is because "science" makes it so! There are many different sciences, each with their own methodology. Why has "ghost hunting" focused on a physical science perspective, almost to exclusivity? Shouldn't "presence" (as a "form of life"), and revealed in social interaction, become the "subject", rather than "object" of inquiry?

The simplified division of "subject" and "object" does not capture the reality of "forms of life" or the complexities of fieldwork, except as a form of <u>superficial</u> entertainment. What should occur in fieldwork is a mutual interlocking exchange of information that is <u>recognized</u> by <u>both</u> "subjects" (investigator and "ghost"), and become a reflection as to the form, timing, and substance of a manifestation.

In this way, the "manifestation" changes from an "object" that is recorded and measured to a "subject" that is <u>communicating</u>. The perception of this communication must not decay into an "object" of jubilation, which it does so often in "ghost hunting". With this celebration (seen so often in "paranormal reality TV"), as the "close-up" expression, the "build-up", the "cut" to commercial, and the "reveal", the fieldwork becomes an "unreal" investigative process.

If a "subject" is communicating (as opposed to an "object" manifestation), one must continue on with the interaction, not <u>end</u> it by congratulating oneself! In fieldwork, by dividing the investigation of past presence into a "subject-object" differentiation, we populate the past with a particular way of perceiving its "ghosts" that is <u>modern</u>, not past. This is imposing our contemporary views onto the past.

This is not an arbitrary vision of past presence. Unfortunately, it is part of the way among many "ghost hunters" in which modernity and its science has divided-up the world as reality. Most "ghost hunting" is not a visionary attempt to offer meaning to what occurs at haunted locations. It is a "dead-end", without learning, and relying on its "entertaining" nature to hide the problems inherent in its methodology!

That is why we need to <u>learn</u> from our mistakes. I have. Are you willing?

"It was a mistake to imagine the past simply buried underground. There was that element, yes, but it might be more accurate to think of it living, breathing, and walking upon the earth as well...."

- Erin Hart, <u>Haunted Ground</u> (2003)

"No space ever vanishes utterly, leaving no trace".

- Henri Lefebvre, <u>The Production of Space</u>

The Landscapes of Learning

Landscapes, especially haunted ones, are produced in tense interactions with other cultural expressions. What results are "ghosts", made tangible through the spatialization of the memory of these occupations of past presences. Multiple agencies of connection turn places into expressions of "extroverted" manifestations. They create residual presences that weave in and out of absence, caught momentarily (for the most part), by an "eyewitness" or "earwitness".

Today we struggle to understand what may be occurring in haunted space. Because we have failed to learn who does manifest there, and what is <u>never</u> there, we live, research, and investigate as "captives" of our own times and beliefs, science included. We are unable to know the past and its continuing presence today. The fault is of our own making, not that of science or technology.

There is a <u>ritual</u> nature to haunting phenomenon. It produces, time and again, <u>expectable</u> behaviors, recognizable if only one thoroughly researches a site's social and spatial histories. This research is not for entertainment purposes, so it requires more than a superficial glance, or an occasional "salute" to history. That we have not learned this yet is our own fault, and is the reason why ghost research, despite increased use of technological devices in the field, has not advanced our understanding of a haunting.

A haunted site, like a "portal", becomes a door through which a complex field of practices and memories enter and <u>remain</u>. These "fields" produce haunted space, peopled by the <u>living</u> in <u>absence</u>, and the "dead" in resident presence. Together, they constitute a stratigraphy of belonging, and the uncertainty of not knowing who is a living presence, and who is the "ghost" of past presence?

A place, with spaces of absence and presence, is what Gordillo (2004) identifies as **"the result of the social contradictions embedded in them" (2004:5)**. This is an appropriate venue for ghost research, for such places reveal **"the fractures....that make them ongoing, unstable, and unfinished historical processes....(Ibid: 5)**. They become active sites to learn about what haunts them.

At these haunted sites, the process of documentation is <u>not</u> easy. In my four decades of fieldwork, I have learned how true this really is! And it's easier to prove what we don't record and experience at a haunted location. It's harder to admit that! And it's still <u>more</u> difficult to prove that this absence of data <u>is</u> significant beyond the idea that the location may <u>not</u> be haunted!

Sometimes, "silence" and "inactivity" is important because it was so as part of past history, and particular spaces in time. A good example of this was (is) Eastern State Penitentiary in Philadelphia (in particular cell blocks), and such was the case with our eight day "embedded" investigation of the Knickerbocker Hotel in Linesville, Pennsylvania (see below).

The concept of ritual, as expectant behaviors in a haunting, could provide a key to why some spaces are haunted. In ghost research, we must learn the power of spiritual belief, and how it can shape an entire socio-cultural world, and way of life, in history. Such was the case of the "good death", its specific rituals of expectant behaviors, and how the Civil War battlefield literally destroyed those rituals.

It led, I propose, to haunting behaviors on many Civil War battlefields. At one such battlefield, Burnside Bridge (Antietam battlefield), Maryland, we learned how, when, and where "ghosts" manifest because they (the "ghosts") learned through recognition and recall from their own past ritualistic behaviors, and beliefs in this "good death" (see below).

Then, there are the haunted institutions, such as schools, prisons, and museums. Architecture, décor, dress, food – all of it – were specifically planned to play a proactive role in accomplishing the institutions goals and purposes. Because of this "institutionalization" of goal and purpose, they can serve as "triggers" today in fieldwork at these institutions.

But we must continue to ask: "What more can we learn"? This requires a holistic approach that explores what individuals (and groups) did in each space/situation in these institutions. We must ask ourselves when and how they did it, and what they did, with whom, and why! Some data might not be known to history. This becomes our opportunity to "fill-in" those historical "gaps" through ghost research. The process becomes, not an re-enactment or re-construction, but rather a construction to make more accurate these histories at haunted locations, and the "politics" of past purpose and contemporary research goals.

We must learn to be contextual and comparative. We must learn to ask ourselves whose methods (and science) we define as the "ideal" against which to measure fieldwork practices in these institutional hauntings. We must learn to shift focus from binary categories to analysis that is based on other criteria. For example, instead of comparing inmate/staff behavior at a haunted prison, why not base our analysis on age? Let's change the teacher/pupil binary distinction to one based on the sex of the individual. We must learn to ask the following: "How do using categories (teacher/pupil; inmate/staff; guest/owner, etc.) "mask" or "hide" what can be "unearthed" of manifesting presence at a haunted location?

I will discuss three such institutions, as learning centers, in this book. These include:

- The institution of hospitality, and our "ghost excavations" at the Knickerbocker Hotel in Linesville, Pennsylvania;
- The penal institution and how to "connect", based on "ghost excavations" at Eastern State Penitentiary in Philadelphia, Pennsylvania; and
- The institution as museum and the "ghost excavation" that we enacted at the Brunswick Railroad Museum in Brunswick, Maryland.

The "Knick" in Time?

"Our sense of the rightful possession of a place depends in part upon our sense of the ghosts that possess it, and the connections of different people to those ghosts...."

- **Michael Mayerfeld Bell**

"Ethnography is a process of creating and representing knowledge....that is based on the ethnographer's own experiences.

- **(Pink 2007:22)**

The "excavation" of the "Knick" is about a host (the hotel), its "ghosts" as guests, and the "witnesses" to that hosting of "ghosts". It is not a complete reality of this presence, nor an exhaustive list of "witnesses", nor is it meant to be. To understand the scale of this interaction, a compromise becomes necessary. It is the study of our "excavations" there, and how we came to learn the nature of what "haunts" the hotel today.

The "Knick", we thought, like all spaces once occupied by humans, should contain embedded signature traces, reflecting the life and times of those individuals who have entered the confines of the hotel, and who participated in activities there. In certain of its many spaces, these traces accumulate, registering the signs of regular and habitual acts that reflect the ordinary and mundane operation of a hotel's many functions to the world at large. The existence of these presences constitute, we are proposing, an ongoing archaeological record of the occupation and social functioning of the hotel throughout its history.

In the habitation and service areas of the hotel, there remain, we are theorizing, authentic traces of the performances of everyday hotel life, including residual elements and interactive traces of social tradition and ritual, habitual acts, and vestiges of accidents, events, and death that are recorded in historical narratives, some of which continue as manifesting expressions in particular spaces of the hotel's three floors. They would represent complex ethnographic stratigraphies of what and who remain of the hotels past cultural worlds. Décor and structural changes through time may alter appearances, but not the vision of presence that may remain there.

Photos 7-11: The exterior and interior shots of the Knickerbocker Hotel

Photo 7:

Photo 8:

Photo 9:

Photo 10:

Photo 11:

The stories that remain, and the "afterlife" histories that may continue, do not cease or pause because the hotel has been removed from the "life" of the people who once lived, visited, and worked there. If there are "ghosts" at the "Knick", then "life" goes on as it once did, and any contemporary exploration of their presences must treat the present as if it were the past.

The hotel and its "ghosts" are constantly in motion, as this is no sterile museum enclosed in glass exhibits. We must not be influenced by the structural and decorative changes (nor its "ruination" in places), as this conditions how we observe, perform, record, and explore the site. We must remain true to the "atmosphere" of the hotel's past, treating the spaces within as they once functioned.

The exploration, "excavation", and documentation of the hotels uncertain past presences must involve a process of cultural production. This is an active apprehension of, and sensitivity toward, the traces that remain. Our investigative performances must demonstrate the context of what (and "who") may remain. Our acts must be redemptive, moral, humane, and therapeutic. They cannot be demanding. They should not command, except in appropriate contextual cultural situations that are warranted by what occurred there in the past in particular spaces. A "new order", going beyond a simple "ghost hunt" mentality, offers new possibilities to connect with the past.

We began the "ghost excavation" of the Knickerbocker Hotel in Linesville, Pennsylvania thinking to emplace ourselves in the spaces of the hotel as ethnographers, recording and documenting the continuing cultural life of the hotel's past guests, staff, and occupants. We believed that our empathetic acts of immersive contextual behaviors, as a form of cultural resonance, allowed us to make that "connection" mentioned by Bell in the above quote. We hoped that this permitted us to record the "apparitional experiences" of the cultural hauntings that have been perceived at the hotel.

To become knowledgeable of this continued past occupancy, we felt that we must go beyond simple contemporary "measurements", through electronic devices, of that continuing past reality. The ethnographic immersion, using contextual cultural scenarios, offered us, we believed, a sense of the haunted character of particular spaces of the hotel. This provided us with a horizontal delineation of haunted space (particular rooms = specific haunting behaviors), with a vertical dimension of historical haunting uncertainties relative to particular situations, events, and acts in the hotel's history.

Using ethnographic process, as described by Pink (2007) above, we attempted to record "apparitional experiences" as **"versions of ethnographer's experiences of reality that are as loyal as possible to the context through which the knowledge was produced" (Pink 2007:22)**. This ethnographic fieldwork throughout the hotel is what Karen O'Reilly (2005) calls:

"iterative-inductive research (that evolves in design through the study).....involving direct and sustained contact with human agents, within the context of their daily lives....watching what happens, listening to what is said...." (2005:3).

This is ethnographic participant-observer fieldwork, done by "living" with the people (in this case the "ghosts") being studied. Limitations are a concern here, as our fieldwork in the hotel only lasted eight consecutive days, and was preceded by 3 other investigations at the hotel, each limited to about 12 hours.

This has meant that innovative methods had to be developed to provide ways into understanding the haunted nature of the hotel, and its occupants. This involved "sharing" activities and practices with the "ghosts", and "inviting" them to enact with us various forms of expressing themselves. It also meant that this present work remains only a summary of our investigative work there, focusing on what we learned during this ethnographic immersion, and how it affects our approach to fieldwork at haunted locations.

This innovative method uses what has been referred to as **"extended science" (Braud & Anderson 1998)**, which also includes an analysis of the researcher's own experiences, linking it to Pink's definition of ethnographic fieldwork. The conditions for doing "extended science" are consistent with the requirements for valid data accumulation in scientific inquiry (cf. Wilber 1983). These include:

- **"instrumental injunction":** such that **"if you want to know this, do this...."**;
- **"intuitive apprehension: This is....the immediate experience...addressed by the injunction; i.e. the immediate data-apprehension;**
- **Communal confirmation: This is a checking of results....with others who have adequately completed the injunctive and apprehensive strands" (1983:40).**

In our ethnographic immersions at the "Knick", these three strands of "extended science" correspond to:

- "Instrumental injunction" = our contextual cultural scenarios in specific spaces of the hotel. This involves specific historical acts, situations, and behaviors, verified by historical narratives about the hotel;

- "Intuitive apprehension" = the "immediate field reveals" as perceived and recorded by the team on our RT-EVP devices, and the continuation of the scenario <u>after</u> a manifestation has occurred; and
- "Communal confirmation" = the "post-apparitional experience" with the team to confirm what had just occurred.

This technique of data acquisition must be completed in its <u>entirety</u> to be used for confirmation purposes according to the proponents of "extended science". If not, the data is <u>not</u> valid.

The technique is both scientific and ethnographic. It is scientific because, as Wilbur (1983) defines science, it is a disciplinary approach **"that conscientiously follows the three strands of data accumulation and verification...."** **(1983:62)**. It is ethnographic because it is an immersion into the past cultural history of the hotel, an immersion based on historical factual data.

In order to use the ethnographic immersion as "extended science", we needed a "cultural map" of the hotel in which to emplace ourselves during the investigation of its perceived haunted space. Concerning a "cultural map", Bourdieu (1977) states the following:

"....culture is sometimes described as a map; it is the analogy which occurs to an outsider who has to find his way around in a foreign landscape and who compensates for his lack of practical mastery....by the use of a model of all possible routes" (1977:2).

A haunted location <u>is</u> a "foreign landscape". Without a "map" to use as a guide, a typical "ghost hunt" summons to the use of many different ideas, techniques, philosophies, and personal preferences. <u>That</u> diversity, without reason or baseline, is what is wrong with contemporary ghost research!

A "map" is needed to systematically explore, not "hunt", for a way through haunted space. I believe this "map" is <u>cultural</u> in nature, and <u>not </u> a physical measurement. The <u>guiding</u> principle that one uses to explore this uncertain space is how a sense of a space is placed, and how it becomes how we sense its haunted character.

This sense of place makes sense through <u>culture</u>, both present acts (that resonate) and past manifestations (that respond to that resonance). Our cultural sense differs from the past, <u>each</u> past at a haunted location. This is not because the past is "dead and buried", but because it <u>remains</u> (as it haunts us) as it <u>was.</u>

The development of this "cultural map" must provide meaning to the sensory images that manifest there in a particular space. The baseline for this "cultural map" must <u>not</u> be visual. It must be historical, a history of the functionality of the hotel's spaces. This is because if a space is still occupied by past presence, then it becomes part of the archaeological record of the hotel's occupancy.

This creates the ethno-archaeological frame for the cultural map of the hotel. I call this frame the "Knick S.C.A.M.". This includes:

- **"S"** or the **Strata**: This is the historical function of a particular room in the hotel, and its occupational history. This becomes the site of "excavation" into a particular "haunting uncertainty";
- **"C"** or the **Context**: This is the situated cultural scenario(s) that we enact for each room "strata", and is relative to that room's function;
- **"A"** or the **Association(s)**: This is (are) the trigger(s) that we use during an "excavation" of a particular "strata" and specific "context"; and the
- **"M"** or the **Matrix**. This is (are) the remains that we "unearthed" during the "excavation" of each "strata".

The "S.C.A.M." frame becomes the "haunting occupation episode" or (H.O.E.) of each unit of "excavation" during the investigative process. In a "ghost excavation", we use the "H.O.E." to document the "unearthing" of past presence at a haunted location.

A haunted location should be explored by paying close attention to the vertical nature (or "strata") of "haunting uncertainties" at a site. This vertical nature does not mean that we physically "dig-deep" at a site. It means that there may be various "layers" of haunting phenomenon, all of which are located on the surface of physical space. We must also follow the horizontal extent of that "haunting uncertain layer". This means that we enact cultural scenarios in other spaces of the location that are also associated with that particular "strata" of "uncertainty. That way, an entire "S.C.A.M." can be "excavated" as one episode, and all manifestations will be documented as part of that clearly-defined "S.C.A.M." deposit.

Another of our goals at the "Knick" was to document the auditory ambiance of the hotel as a "sound" stratigraphy of presence. There is ample "room" here for exploring the sonic stories of continued occupations of particular spaces, however fragmented they are, of the former hotel. By listening to the "world" inside the various rooms, we create representations that bring together different voices from those (continuing to be occupied) spaces.

The objective was to hear "what" and "who" remain, and the "experiences" of these individuals as they are encountered in our "excavating" performances, ensuring us an expanded audio version of "actuality" at the hotel. Such an approach is a creative treatment that reflects the sonic remains of occupations at the hotel in a far more interesting way (as a "ethnography of communication" or E.O.C. rather than an EVP) that those simulated in paranormal reality TV "entertainment".

This E.O.C. involves the <u>construction</u> of a sonic environment that becomes representative of continuing past reality. This auditory composition of a haunted hotel is a mix of gathered sounds, voices, and social ambiance that bear witness to an occupied reality. This became an <u>engaged</u> "theatre" of "stage" presence, composed of an inter-locking historical script with emotive elements of real past drama. This past drama reflects the auditory S.C.A.M. of an "archaeological imagination" similar to **Theatre/Archaeology (Pearson and Shanks 2001)**.

This cultural map of the "Knick" as a "theatre" within which we <u>performed</u> our investigative acts is an adequate conceptualization of the spaces of "excavation". However, the original design of this "theatrical" setting, and our emplacement within it, proved inadequate to what occurred during the eight day "excavation" of the hotel. These problems will be discussed later (see below).

Composing this "Knick soundscape" creates an emplaced sense of presence that becomes the construction of reality of the contemporary hotel. The spaces, rooms, and floors that are sensed gives us a sense of place (as "haunted" or not), and this sense of place becomes "streams" or "flows" of what "haunts" the "Knick".

The baseline for this soundscape was a "peripatetic walk", noting the various aural elements of the hotels sonic ambiance. This "sound walk", as a form of portable sound technology (using headphones and a RT-EVP audio recorder), is a more intense sensorial form of "observation" than merely watching electronic measuring devices, viewing a video monitor, or periodically scanning photos during a "sweep" of the site. The walk becomes a triangulation of sound, space, and memory.

An emphasis on the purely visual dimension greatly affects our perception of other sensory modalities. Yet, what we "see" has little or no relation to what we may hear, and what we hear (as intense "listening") makes the visual experience quite monotonous: "empty" space most of the time! Hearing is a focused concentration, and sounds become an intense contact between space and the sense of it. And that is why we concentrate on the aural dimension.

When combined with contextual and resonating scenarios in particular spaces, they "signal" a clear learning process that can be attributable to a particular past presence (the one that is "targeted"). And any "sound" response may be viewed as a recognition of the investigative act as purposeful and communicative to a particular past situation, the recall of that past situation by the "ghost", and an attempt at communication by the "ghost" in a manifesting sonic form.

In this peripatetic walk, however, no distinctions were made between past and present acousmatic sounds (source/origin unknown), nor did we separate (and make a distinction between) "residual" and (possible) interactive "soundings". One lesson that we learned from this "audio walk" is that we must become more specific as to what we are actually recording, and the source of that sound.

Photo 12: The "Audio Walk" at the "Knick"

Our "sound" sense of the "Knick" included sonic elements that went beyond "typical" EVP responses. These included recordings of what Noory and Guiley (2011) called **"paranormal parrots"** and **"spontaneous time displacement" (2011: 137, 141)**. The former is a form of "mimicking". The ghost repeats what they hear, using exact or similar words/meanings. The later involves "tuning-in" **"to an ever present, eternal now where everything is happening simultaneously" (Ibid: 141)**. We have recorded numerous examples of this sonic behavior. Many are responses to similar sonic field patterns in the past, and linked to <u>particular</u> spaces in the hotel.

An example of a "paranormal parrot" at the "Knick" involved the recording of "shush" in the "ski" room on the third floor. Here, the entity would first tell me (personally) to "shush" (be quiet) every time that I said something, and <u>only</u> when I said something. I would "shush" back, and the entity would repeat my "shush". This occurred multiple times. Also, when one of the investigators (Mary Becker) was counting numbers in the hallway to the "children's bedrooms" on the third floor (the former "ballroom"), we recorded the voices of a child repeating the numbered sequence. This also occurred several times. You can hear these audio recordings on our website at www.ghostexcavation.com.

An example of a "spontaneous time displacement" was recorded several times in the rooms and hallways of the Knick. These included the verbal transmissions of investigators, as they enacted scenarios (the present), and the opening and closing of doors in the <u>same</u> space (and visually unobserved) during the scenarios (the past). This occurred several times, both in the day and during evening hours. These "entrances" and "exits" were contextual to the scenarios that we were then enacting (such as bringing "guests" to their rooms; going to the "ballroom to dance"; going to the "Opera" through the narrow "hallway" that was once located on the 2nd floor, etc.). You can also hear these audio tracks on www.ghostexcavation.com.

Interestingly, we have numerous recordings of animal sounds in the hotel. These sounds are principally dog and cat sounds. The sound of a dog barking, recorded on the second floor hallway was contextual. We were doing a scenario of one specific hotel guest who had a dog. While doing the scenario, the dog barked. While preparing for the scenario where we were going to enact the hotel opening "ballroom" dance on the third floor, we heard the sound of a cat "meowing" three different times. The first two sounded like an adult cat, while the last one sounded like a kitten. There was no cat visible to us on the floor where the sounds originated. You can hear these animal sounds on our website at www.ghostexcavation.com.

We also recorded several "transductive" sounds, i.e. those that were unanticipated and which transcended spatial and temporal parameters. An excellent example of this was the voice of a child who not only "played" a "counting game" with one of our investigators (Mary Becker), but who <u>also</u> "voiced" an <u>independent</u> and personal expression that taught us that this "sounding" did <u>not</u> "mimic" Mary's voice. It went <u>beyond</u> the "game" to a previously-learned numerical sequence: as Mary was counting, the "child" began to count <u>backwards!</u> We also have a recording of a child, while Mary was "counting", repeat the number "5" several times, independent of the numerical sequence. You can also hear these recordings at www.ghostexcavation.com.

Photo 13: Mary during the "Counting" Game

In their book, Noory and Guiley raise some questions about the actual identities of these EVP responses:

"...how confident can we be that we are talking to the dead or other spirits? We may be tuning into minds somewhere on the planet – or perhaps in a parallel dimension (2011:143).

My answer to those questions is that, during our "ghost excavation, we had established <u>context.</u> Those audio recordings all were recorded (except for the "transductive" ones) within the particular scenario that we were enacting in a specific space. Context is the key, I propose, that can unlock the answers to their questions.

This context was established in the following manner (as our investigative sequence of acts):

- The "out of frame" (haunted space) preparation. Here, we discussed the various roles of the investigators as "performers" in the scenario. Each role and scenario was discussed thoroughly. We also talked about the documentation "set-up" (where to place the cameras, audio equipment);
- "set-up" video and audio recorders "in frame";
- Execute "dead-time" out of frame for about 10 minutes. This also included "mental" preparation (for "empathy");

- Perform scenario "in frame" by first establishing our "identity" as a past participant in a particular contextual situation (one that resonated with what may have occurred in a specific hotel room/space);
- When a "manifestation" occurred (in any sensory mode), continue the communication based on what had just occurred;
- If nothing occurs, or if "manifestations" stop, end the scenario and walk out of frame (after about five minutes duration); and
- Discuss the scenario out of frame, making adjustments (if warranted) for future iterations of the scenario, or something similar (contextual and resonating).

This sequence of investigative acts and field process was performed for each and every scenario that we enacted during the "ghost excavation" at the hotel.

While enacting these scenarios, we learned how to adjust our investigative stance in order to better get into the flow and rhythm of the hotel as a hotel, not a haunted location. We noticed that days were usually "inactive" ("as guests leave"), and nights were more active ("as guests arrived or returned"). During the day, we performed as "household staff", cleaning and arranging the rooms, or doing "simulated" maintenance. At night, other spaces became more "active", such as the former restaurant/bar, and the "ballroom". In each of these areas, we performed several resonating acts.

In this way, we felt that these acts increased the possibility of predictability as a consequence of the resonance to past interaction, behavior, and activities. This assumed that the "ghosts" as remaining "guests" or "visitors" would recognize the situation as enacted, and us as "one of them" in a past activity. Their "manifesting" presence would, I propose, be a "recall" of appropriate behaviors, as a pre-established pattern or habit that is remembered.

Photo 14: The Performance of a Scenario at the Hotel

Photo 15: The Performance of a Scenario at the Hotel

Photo 16: The Performance of a Scenario at the Hotel

Our embedded stay at the hotel coincided with the time frame of the original opening of the hotel (January), and we enacted various scenarios relative to that "opening". These included:

- An "interview" with the original owner, Milo Arnold. The conversation was based on the newspaper description of the hotel's opening;
- The arrival of the first guests;
- The dinner and reception in honor of the "opening";
- A married couple's incident, as described in the newspaper account; and
- The ballroom dancing with period music.

We enacted several scenarios in the area of the ballroom on the third floor (later the Arnold residence), as this area was considered the most "active" area in the former hotel. We also enacted several scenarios relative to the original guests of the hotel when it first opened, focusing on their "occupations" as stated in the 1880 census.

We enacted several scenarios relative to Clara Arnold's fatal illness and eventual death due to consumption. We thought that her obituary and the announcement of her death, as placed in the local newspaper, afforded us some historical material in which to enact several scenarios.

We sometimes enacted two different scenarios simultaneously, in two different spaces and floors of the hotel, to record any movement and manifestations in the "between the scenarios" spaces. Most of our scenarios were based on the historical record, but the reasons for actions and the motivations of individuals were not known. Because of this, we "acted-out" several possible situations for these scenarios.

During the performances of these scenarios, during the seven days that we stayed at the hotel, we learned that the execution of this methodology (contextual/resonating scenarios in specific spaces of the hotel) was somewhat flawed. And our concept of a "theatrical stage" which "targeted" specific situations and aimed at particular occupants and staff of the hotel (in a particular layer of "historical uncertainty") was too restrictive as an "excavating" procedure.

Using the concept of fieldwork as "theatre", and our "excavation" space as "center stage", we sought to direct our scenarios (as "cultural scenes") directly at an "audience" in front of us (particular layer of haunting uncertainty). We found that, sometimes, the layer of history and individual that we were directing our efforts towards were not the only presences that were manifesting. And there were times that we had a manifestation that was completely unrelated to the scenario that we were using. The manifestation of "cat sounds" on the third floor, during a scenario relative to the ballroom dancing, is one example of this. When we were enacting a scenario in one of the children's bedrooms on the third floor, one of our investigators had a distinct impression of a man in the room, perceived as "dressed" in the period of the 1940's (we were "targeting" the period of the 1880's). During the seven days that we spent at the hotel, we recorded other instances of these multiple presences, or the manifestation of "others" that were "out of time" from the investigative acts that we were enacting.

Did the hotel décor contribute to the appearance of these "other" manifestations? Do the multiple and different room decors attract wandering entities, in the form of a "portal"? Or, do our engaged scenarios attract an "audience" of "permanent guests", even though we were not "targeting" that particular time period? Does either of these possibilities affect the concept of "memory" as that which survives as part of an "afterlife consciousness"? These are questions that we have to learn about through continued fieldwork at the hotel. We have learned much from our numerous "excavations" at the hotel, but we have <u>also</u> learned that there is still much work to be done on the "Knick" haunting, and its implications for ghost research!

"The old house, for those who know how to listen is a sort of geometry of echoes....the inflections of beloved voices now silent"

- **Gaston Bachelard**

The "Knick", unlike the "old house" is not always silent. There are "echoes" there, yes. It is the resonance of each room, and its guests, as a hotel of sound. This "sound" map speaks for the history of the place as comfort zones and/or spaces of temporary distractions, be they visual, auditory, gustatory, tactile, or olfactory. Each sense plays different roles within the spatial history of the hotel. These "roles" become the memory of the hotel, sketched as sensorium.

One role that the "Knick" has not "played" is certain mythological ones, and some legendary film lore. In ancient Chinese mythology, an abandoned musical instrument is a symbol of social decay and/or old customs in decline. This is certainly not the case with the décor of the rooms on the 2nd and 3rd floors. These rooms are meant to resonate with particular periods of history that parallel the location's history as a functioning hotel.

It has been said that the piano is most conductive to ghostly activation, and that objects, normally silent, are depositories of the "soul" of a building. Their "auditory" activation indicates that the building itself has chosen to speak. This type of haunting can be seen in such films as **The Innocents (1962)** and **The Changeling (1981)**, among others. With all the possibilities of musical events, the "Knick" itself did not "speak" to us. And there were many opportunities: piano, vitrolas, radios, etc. in the various rooms. Apparently, mythology and film lore did not "translate", at least for us. Does this signify that the "Knick haunting" involves people (who lived, worked, and died there) rather than the building itself?

A sound from another place, seemingly "out of place", is normal for a haunted hotel, as most occupations are of a transitory nature. This complicates the mapping of sounds as to their "origin", but not their haunted nature. There, one also records the sometimes anomalous sounds coming from creatures of habit, "non-human" in origin. These are animal or insect, and which further disturbs the "restful" soundscape.

How do such disturbances violate the subtle audio flow of ghostly presence? These "creatures" too are tenants, but are they "ghostly" ones? In darkened corridors and unlit rooms, sounds amplify visual absence and rational thought. Sometimes fear arises and attaches itself to the fieldwork as one hears sounds which are activated by these unseen presences. They "play-out" interacting among the rooms, the floorboards, creaking stairs, the dark, the prolonged silences, and the investigator as vulnerable listener who is "witnessing" no one.

As the sounds fade, even though recorded, their "reputation" as documentation diminishes becoming at best an unreliable "witness" to past presence. Was that sound an auditory "ghost", physical noise, a memory residue, or nothing at all? At the "Knick", we learn, like the specters that inhabit the spaces of the hotel, that we still have "time" to find out and document which "phantom" sound is a "ghostly" one, and why!

The Lessons Learned From a Civil War Battlefield

There was one who was there, who saw and heard the battle as it was fought. There is one who still lives who witnessed those events of September 17, 1862 at Burnside Bridge, Antietam Battlefield (Md.)

Photo 17: The "Witness Tree" at Burnside Bridge

But I am human, not a "witness tree". To learn about the Civil War, I did not merely go and read the historical narratives of the battle, or the re-constructed histories of modern historians. That would have been too easy, and the most logical path to take for an academic study. Instead, I chose to seek answers on the "fringe", away from any "borders" that would enclose my mind or thoughts.

To seek some answers, I chose to "perform" on a Civil War battlefield. I chose to record "responses" to those "investigative performances" as someone whose "presence" <u>was</u> significant to what remains of the past at Burnside Bridge. And, in the process, I re-learned what <u>more</u> still remains of the Civil War that goes beyond contemporary thought, analysis, and reality.

The direction of my performance and its exploration was not measured by scientific instruments or academic approval. It was learned by recalling what it is to be human, and immersed in some horrific situation. Though I was not involved in deadly consequences as those soldiers who fought there, I was sympathetic to their dreadful circumstances:

"In order to go forward, one has to go back.

To go back, one has to turn around.

To turn around, one has to pull over and look to

see if anyone was coming"

- **Linda Mussman, "Civil War Chronicles" (1988)**

When the fieldwork begins anew, they all will be waiting. In haunted battlefield spaces, we must first recognize what we are really experiencing and recording in measured tones. We must recall that struggle for space, and "ground" our thoughts on "who" may be coming present to meet us. At the exchange, we must respect both them, and ourselves. This means to not fail to <u>remain</u> human, and to treat these presences as humans who are advancing once again over contested ground where we now tread in peace.

"Death appears to result in the paradoxical production of both disappearance and remains. Disappearance......clings to remains – absent flesh does ghost bones" (Schneider 2011:103).

We must "dig-out" this haunting to distinguish absence from presence.

A battlefield "haunting" is thrice-behaved behavior of the "culture of war" (of any war). This behavior is:

- <u>"Composed"</u> in drills;
- <u>"Practiced"</u>on the battlefield; and
- <u>"Remembered"</u> in a "haunting".

This is "C.P.R." for "ghost excavations" on haunted Civil War battlefields. During a "ghost excavation", our "investigative presence" becomes a "strong" performance which, in some instances, is <u>recognized</u> by the "battlefield ghost" because it corresponds to some behavioral act of the "culture of war". It is something that these "ghosts" learned by composition, practiced, and still remember.

They behaved on a battlefield as they were trained to behave, based on the learned practices of what it meant, and how it was to fight in combat during the Civil War. The physical acts of combat of the "theatre" of war followed a learning precedent, and left "remains" for others to follow the experience of "seeing the elephant".

A "ghost excavation" is "trench warfare". "Remains" of past presences are "dug-out" in the spaces between times, between fields of recognition and recall, and between contextual and resonating acts of investigator and "ghost". Battlefield movement, once obscured by smoke, requires a reiteration of "re-sounding" acts implied in "digging deeper" on the surface of that which continues as a "learned" haunting.

A "ghost excavation" field <u>performance</u> differs from "ghost hunting", and its (usually) simplistic acts of scan, measure, record, and monitor. A "ghost excavation" involves a goal-directed, intentional activity, using some model in that intention to predict manifesting past behaviors. For example, at Burnside Bridge, this model was the Inherent Military Probability (I.M.P.) behaviors (or what the soldier would have done and experienced in particular situations of certain space) of the "culture of war" of the American Civil War.

The true "spirit" of the "ghost excavation", its purpose, is to interpret a past situation and sentiment. This does not make the experience "original" again. It is not meant to replicate past acts, events, or situations. And it does not "repeat" the past by commemorating it! One does not become "overrun" emotionally by the past (a "wargasm"). One does not, either, go to the other extreme, and have a "ghostgasm".

We learn to remain "liminal", betwixt past and present (an "untimely" situation), and between a participant and one who also observes. We remain "grounded", in that space of "excavation". We are not the "cultural informants", as some hard-core re-enactors believe they become. The battlefield "ghost" is a "phantom" because re-enactment makes it so, as a sense of the past as past, though not gone. It remains "available" to "re-do" or to "return to". This is not the same as "unearthing" the past in a "ghost excavation". A re-enactment is a repeat of the past. A "ghost excavation" is learning from "what" and "who" still remains from that past.

What we strive for in a "ghost excavation" is "affect", not the "authenticity" sought after by serious re-enactors. This "affect" is appropriate because it arises in the midst of in-between-ness, our "liminal" position in haunted space. This is an investigative stance between doing, and learning from what we do, its "affect". The effect provides us with a key to a world of <u>potential</u> encounters, and is <u>not</u> "copying", no matter how "authentically" it is performed, that past.

The "affect" of battle for all Civil War soldiers was a sense of the <u>past</u>, as something learned in drill. For some, it became a <u>future</u> direction, a "haunting". This "haunting" aspect of a Civil War battlefield can be re-discovered (or "re-called") as something familiar through a resonating and "strong" investigative "presence" that "affects" its <u>recognition</u> by that "ghost" on the battlefield. It is the traumatized soldier (as "ghost") who re-lives a battle in the <u>future present</u>, due to the shock of death.

A battlefield "haunting" is an <u>affective</u> present "present time" experience of war, beginning again through <u>effective</u> investigative presence. This "effective" presence prompts "habit memory" (of drills) of the "culture of war", and turns it into the recognition of these same habits on the contemporary battlefield.

It is this "affect" that we sought to document during our fieldwork at Burnside Bridge. We asked ourselves: what will be the "affect" of our investigative performances, and why would it "effect" those who may remain there from the past? If there was an "effect", what can we learn from them?

The haunted nature of this history and that "something" <u>still</u> remaining returns us to this transmission of "affect". At Burnside Bridge, you don't feel like a "witness" because there is little or nothing to see. We became "earwitnesses" to actions we clearly did not see, a <u>clear</u> resonance to what these soldiers experienced on that battlefield in 1862. Then, as now, Burnside Bridge, as a "haunting" space, was (is) not a landscape. It was, and remains, a "soundscape" of what occurred there in 1862. This event of battle is given to us, but it does <u>not</u> concern our ability to see it. It takes place amid the "fog of war", and is sensed as such!

We don't see the world beyond. We hear it. And I am not "sounding out" entirely from sounds that emanate from a separate consciousness, or belonging to a specific <u>identifiable</u> and visible space. There are times when the "unusual" becomes "normal". The "norm", however, is <u>not</u> to accept this. Yet, it occurs "normally" (and quite uneventfully) when we focus on listening, and not merely hearing what we want to hear. There is ethnographic precedent to the idea that sounds that manifest in particular locations are the vocalizations of spirits or ghostly entities (cf. Goldman 2003).

Some sounds outreach the known and the world we negotiate through sight. What happened to insight in the equation of exploration? Let us venture beyond visual comprehension, and "beyond the paranormal".

Let us reach a vision of reality by hearing what is "below the radar". There are indeed times when the behavioral space of sound expands after physical death, as "soundings" extend into the future from the past. In an age of mass media, this is not difficult to "see". It is difficult, however, to "view" these sounds as something "normal", a part of the continually expanding reality of the world.

A "sound" world inhabits and "lives" on American Civil War battlefields. Despite much "heard" silence today, sounds accumulate in battlefield spaces. They travel beyond the sounds of silence between the "soundings" of historical narratives, beyond the once impenetrable division between North and South, and the contemporary historian's dividing sentences of explanation.

The "unsaid", in these works and words, become "heard" (not "seen" in books). During the Civil War, the non-witnesses to battle scenes only <u>saw</u> the <u>aftermath</u> of these bloody engagements in photographs. They did not <u>hear</u> the battles (for the most part), and only then from a distance of safety. Sounds were described and imagined in paintings, line drawings, and first-hand accounts. But, it is only today that we can <u>experience</u> those battles with the men who fought them, through the sensory modality of sound.

Notions of "authenticity" were built onto these photographs, illustrations, first-hand accounts, and the historian's world of battle. This "evidence", tracked through written records and oral histories, is an incomplete history lesson, and an unbalanced vision of reality, as material source and agency.

To prove the existence of a "ghost", it does not necessarily mean we must recover the "corpse", a "full-bodied apparition" (or a "ghost hunting orb"). This is a prejudicial (over-emphasized) and narrow-minded approach to gathering data. It illustrates the deficiencies of a rigid sensory hierarchy that establishes the visual as the most important sense to make "sense" of a haunting.

This equates the visual with knowledge, veracity, and evidence. The vision of ghost research (as "ghost hunting") is seen as an obtainable "sighted" goal, and the primary interpreter of an "apparitional experience". The "ghost" as a particular "formed" presence haunts the world of ghost research. It is an ill-informed "object"-placed emphasis on "sightings". This was not the world or the "ghost culture" (Sabol 2008) that we encountered and heard at Burnside Bridge.

Today, through fieldwork in ghost research on these haunted battlefields, we can document an expanded sensory world of the men who fought there, and their experience of battle, its suffering and deadly consequences. This _is_ a "normal" exploration of historical events, and of the continued social reality of the "culture of war" of the American Civil War. We are now "hearing" what _they_ heard, and their experiences in these "battlefield soundscapes".

It must be emphasized that what we are recording on these battlefields of what they heard _is,_ nonetheless, biased. As Toop (2010) says:

"The further back in time we travel, the less certainty can be ascribed to the way people once listened, what it was they heard, and what it was they believed they heard" (2010:16).

This doesn't make what we record any less real of how it might have sounded, at the cost of making it _more_ "paranormal"!

In that respect, we have a distinct advantage over those who fought in battle. We have recorded both the residual elements of a battlefield soundscape, and the interactive experiences of some men as they "vocalized" them to us during a "ghost excavation". Unfortunately, we cannot record the sounds of their blood spilling as they died, or the sounds of their thoughts as they left the physicality of combat. We _have_ recorded their "life", however fragmented it is, in battle (and not their death). But we still have much to learn from that recorded sound experience.

Our exploration of the Burnside Bridge soundscape is a learning tool, part of a sensory tool-kit for ghost research. It is meant to re-define the contemporary battlefield space, "calls" into question states of being and being present, and alters a time base that has always been considered "dead" and past. The "excavation" adds depth to our experience of place, one which requires "digging-deep" to learn its sonic (and once silent) secrets. Sound, and its recording, becomes a highly significant component of the cultural map of the contemporary Civil War battlefield setting at Burnside Bridge.

Sounds create and document an event. They are time sensitive. What we recorded at Burnside Bridge "sounds" like what occurred there on September 17, 1862. A space, such as the Burnside Bridge landscape changes appearance, but the sound there remains compatible to who spoke there, or what produced those sounds in that landscape. These sounds remain far longer than many elements of agency that originally produced them. The guns do not function, but the sound remains. The object was not just to record this sonic existence. It was to document its manifestations relative to contemporary investigative acts. This changes a chance residual recording to intentional interaction. This intentionality or purpose of presence was what we sought to document at Burnside Bridge.

What did we learn from recording these sounds at Burnside Bridge? What did we learn from the experience? Was it description, commentary, dialogue, someone singing as a prelude to "battle trance" (Jordania 2011), the sounds of men in battle, was it all of these or mere noise, coincidence, or confirmation bias? Did it become a "map" of memory or merely charting expectations, compromising data for confirming belief? Are the sounds the invisible traces of memories that have been collected here before, during, and after what occurred on September 17, 1862? Just what was the significance of "excavating" and "unearthing" these sounds? There are many questions. The answers lie on the surface of what we gathered at Burnside Bridge, in the soundscape that still "haunts" the present there.

Photo 18: Recording the Soundscape at Burnside Bridge

Photo 19: Recording the Soundscape at Burnside Bridge

An American Civil War battlefield was primarily a soundscape, not a landscape, for the common "foot" soldier. The intense (and blinding) firepower that was generated onto the environment, in mostly restricted spaces, obscured the vision of the landscape setting.

Photo 20: The Burnside Bridge Landscape as Restricted Space

This battlefield soundscape was tied to particular "external experiences" (hearing specific soundmarks in particular spaces/during specific times). This "audio/vision" of an obscured landscape prompted a specific "inner state" which Jordania (2010) has termed "battle trance". This "battle trance", I propose, created specific cultural and mental "fields" (cf. Sheldrake 2012) of the culture of war" of the American Civil War, imprinting these "fields" onto the physical environment.

As part of the "culture of war", these soldiers developed their sonic abilities to know the "external experience" of combat on a Civil War battlefield. This created, through repetition, a sociocultural "tradition" that involved a sensibility and sensitivity to particular contextual sounds or "soundmarks" that were recognized by the soldier in combat situations.

Did this "imprinted" experience become patterned and last after the death of the physical body? Does it become a fundamental social and mental "field" of an "afterlife conscious mind" that survives today at Burnside Bridge? Does it form an historical pattern of individual (and collective) social habits that remain as vestiges and traces of the "culture of war"? If so, do these "fields" become expectations (and manifestations) of what it is to remain human in a given time and place from a particular time and place?

I suggest that the auditory streams that we have recorded during extensive fieldwork at Burnside Bridge might indicate the survival of some form of social/mental "field" as patterned acoustemological presences of this "culture of war". If this acoustemological modality, as a sensorial battlefield "external experience, did become encoded as a "field" pattern, then it should materialize as a unique auditory repertoire and configuration of the soundscape in particular battlefield spaces. This additional acoustical element should, to use Steven Feld's terminology, **"lift-up over"** the **"soundings"** of contemporary sonic elements and vocalizations, and be contextual to the "soundmarks" and auditory streams that would have occurred on the battlefield (in particular spaces) in 1862.

These "fields" (as behavioral "acts") were first developed by habitual drills off the battlefield. They were re-established and imprinted on the battlefield soundscape by auditory cues which repeated the sounds and behaviors of this habitual drilling. These "fields", as memory "tracks", surfaced in battle and produced Inherent Military Probability (I.M.P.) behaviors, or what the soldier would have done in particular situations on the battlefield. It is these cultural (I.M.P. behaviors) and mental ("battle trance") "fields" of the "culture of war" (as "inner states" and "external experiences" in combat) that survive, I propose, as "forms of life" of "afterlife conscious minds" on these battlefields.

These cultural and mental "fields", learned from habits ("drilling"), and reinforced on the battlefield through sensory cues ("soundmarks"), begot a relationship between the acquisition of cultural knowledge (I.M.P. behaviors of the "culture of war") and a focus on a particular human sensory modality (auditory). This acoustemological sense of knowing how to act became a natural (albeit a habitually "haunting") process in battle. And it became an essential part in creating a specific human "community" (a Civil War "company" of soldiers) who identified themselves as a "band of brothers" (both literally and figuratively).

The acquisition of these social and mental "fields" involved a process, I propose, of **"self-resonance" (cf. Sheldrake 2012)** in combat, repeating past behaviors that were learned in drills, cued to particular "soundmarks". In each of our various "excavations" at Burnside Bridge, we have encountered (and interacted with) what Sheldrake has termed **evolving habits" (2012:85)** as a kind of memory. These habits, according to Sheldrake, **"grow stronger through repetition" (Ibid: 97)**.

Does a manifestation of past presence at Burnside Bridge become present as a consequence of (and influenced by) what happened before? Did <u>past</u> drills influence <u>future</u> behavior on the battlefield? Do <u>present</u> investigative acts that are culturally-resonant to a particular space and time influence and cause the manifestation of past I.M.P. behaviors? If so, are these manifestations <u>intentional</u> acts of an "afterlife conscious mind" of a Civil War soldier?

This process of habit formation ("evolving habits"), what Sheldrake calls a "morphogenetic field" includes these

- Social fields – which **co-ordinate the behavior of social groups"**; and
- Mental fields – which **"shape the habits of mind" (2012:100)**.

Do these social and mental "fields" survive as interactive traces of the "culture of war" at Burnside Bridge, and do they reflect the behavior of soldiers who died in battle more than 150 years ago?

If a Civil War consciousness survives the physical death of men who fought at Burnside Bridge, and we repeat elements of their "culture of war" during a "ghost excavation", then these "fields" could become increasingly habitual, as resonating behaviors (present to past) are repeated. The recording of these "repeating" manifestations are not "paranormal" events since the pattern (as social/mental "fields") were already present there in the past. This "field pattern" was not beyond "normal" ("paranormal") since it was originally created in drills, and subsequently executed on the battlefield. A "haunting" pattern as "habitual" behavior (social/mental "fields"), one normally absent, becomes actual given the reiteration of appropriate circumstances, the cultural resonance to these past "fields" during our investigative performance practices.

Sheldrake suggests that mental causation, the "field" that shapes the habits of mind, runs from the present into the past (2012:121). Do our field acts, in specific battlefield K.O.C.O.A. spaces, and cued by specific "soundmarks", "trigger" the actualization of the past social and mental "fields" at Burnside Bridge? If this "afterlife conscious mind" survives the physical death of the body, then our audio recordings of the Burnside Bridge soundscape might reflect this. If **"minds choose among possible futures" (Sheldrake 2012:129)**, then these manifestations, coming immediately following our contextual acts, are purposeful past acts that become present as a response to these contemporary cues. They become the habits of past minds that are responses to present acts.

According to Sheldrake, **minds are closely connected to fields that extend beyond brains in space....and....in time, linked to the past by morphic resonance (2012:229)**. These "minds", as an "afterlife consciousness", remain attached to the social "fields" of I.M.P. behaviors of the "culture of war" which remain in contemporary space and time due in part, I propose, to the retention of residual elements that have been recorded onto the Burnside Bridge landscape, and survive today as part of a Civil War soundscape still heard there.

The concept of "normal" at Burnside Bridge becomes the manifestations of past and habitual acts in the present due to this morphic resonance. "Normal", then, is **not** "paranormal". It is what actually happens in the present, and what happens is based on what occurred in the past in particular battlefield spaces (K.O.C.O.A.).

If this hypothesis is correct, can we then predict when, where, and how these manifestations will occur again in the future? If, as philosopher Henri Bergson (1946) states, if memories are direct connections across time, then these manifestations are memories of I.M.P. habits that are "unearthed" during a "ghost excavation". This is because the enacted contemporary scenarios incorporate resonating acts of I.M.P. habits that "cue" past memories that were habitual acts in the past.

If these past memories depend upon morphic resonance, as Sheldrake suggests, then this "afterlife consciousness" (as entities with memories of habitual I.M.P. acts) is influenced by the morphic resonance from their own past as well. This is the "self-resonance" of habitual drills transferred to the battlefield. Thus, an "afterlife conscious mind" is awakened in the future by a similar resonance (our contextual scenarios) that "targets" (during a "ghost excavation"), this already existing past habitual memory.

The manifestation of memory (as remembrance) occurs as a two-fold process (Sheldrake 2012:204-06):

- Recognition: This is a **"similarity between past experience and previous experience"**; and

- <u>Recall:</u> This is an **"active reconstruction of the past on the basis of remembered meanings"**.

Morphic resonance is seen to connect these two processes, and this connection forms the basis for our "ghost excavation" methodology, and its application at Burnside Bridge:

- <u>Recognition:</u> This aided the Civil War soldier in battle. He recognized the "soundmark" cues learned from drills, reacting to them by taking appropriate actions on the battlefield (as I.M.P. behaviors). These remain today as "vestiges" and "traces" of the "culture of war' at Burnside Bridge;
- <u>Recall:</u> This process allowed us to "unearth" this pre-established memory of social and mental "fields" in specific battlefield spaces (the K.O.C.O.A.) by using appropriate behavioral acts (I.M.P.), cued to the playback of particular "soundmarks".

These recall acts were <u>recognized</u> by the "afterlife consciousness" of these remaining entities as <u>learned</u> past acts. They manifested on the contemporary battlefield where they were <u>previously</u> <u>recognized</u> by these soldiers in battle from <u>prior</u> habits learned in drill.

Sheldrake believes that **"self-resonance from an individual's own past is more specific and....more effective" (2012:211)**. In locations where we have recorded "vestiges" ("residuals") of the "culture of war" (such as those we recorded along the Rohrbach farm road), as elements of this "self-resonance", it is possible that they could also have influenced the continuing presence of interactive manifestations there. The recognition of these residual battle sounds (and perhaps related to the topographic features there (water/stone bridge)) could have enabled a recall of I.M.P. behaviors in this (and other) K.O.C.O.A. spaces by those entities who survive as "afterlife conscious minds".

How can manifestations, such as this "afterlife consciousness" and related "residuals", perceived as "ghosts", exist in contemporary reality 150 years after the battle? There is one possibility. Sheldrake states that **"minds are closely connected to fields that extend beyond brains in space....and....in time, linked to the past by morphic resonance" (2012:210).** When these soldiers died, did their "mind" (as a self-organizing system of I.M.P. behaviors) remain on the battlefield, and become "re-animated" by "self-resonance" from their own past and from the future (the "ghost excavation" practices that we enacted there)?

I propose that a mutual learning process (their recognition/our recall) evolved and has transformed the contemporary landscape at Burnside Bridge. This process includes:

- A past recognition of habits established in drill;
- A past recall that is reinforced by the continuing presence of "residuals" of the battle that are recognized by an "afterlife conscious mind";
- A present performance practice that recognizes the importance of enacting resonating acts in specific battlefield spaces (K.O.C.O.A.); and
- A past recognition of these resonating practices by the "afterlife consciousness" of those remaining as being similar acts to those that occurred at Burnside Bridge on September 17, 1862, and in drills that recall these as I.M.P. behaviors of the "culture of war".

Do the increase in the frequency of these manifestations, during subsequent "ghost excavations", attest to a learning process by an "afterlife consciousness"? Does the use of contextual "soundmarks" as "triggers" reinforce this learning process? Do our contextual acts, portraying those in command of troops in particular K.O.C.O.A. spaces, identify us as "instructors" in the principles of I.M.P. behaviors in the "culture of war"? Hendon (2010) states that there is a **close connection between identity and memory, once the one** (identity) **becomes something recreated over time"** **(2010:14).** Has our "identity" been established by continually repeating these contextual acts (and associated "soundmarks") in "ghost excavations" there?

A morphogenetic field has been established, I propose, through this learning and "excavation" process, and the "field" spreads through time and space to other presences and locations at or near Burnside Bridge. We have documented this expansion. This means that these other (subsequent) manifestations may be a result of <u>recognition</u> and <u>recall</u> as an expanding morphogenetic field which further extends the contemporary reality of "what" and "who" may remain at Burnside Bridge as "afterlife consciousness" and various residual elements.

This potentiality of a learning process in the "afterlife" is <u>not</u> "paranormal" or "supernatural", but learning <u>within</u> the framework and cultural traits of what these "ghost soldiers" already know: the "culture of war"! Their manifestations (and the continued expansion of this "haunted" social and mental "field") are not forms of confirmation bias as an expectation of what should occur when resonance is used as an investigative practice. It is a <u>pre-disposition</u>, a form of "self-resonance" from a pre-established "field" laid down in battle, and in prior drilling. You can hear these audio tracks at www.ghostexcavation.com.

Our continuing fieldwork at Burnside Bridge, and other battlefield sites, is an "audiography" of the "culture of war" that remain as "vestiges" and "traces" of battlefield soundscapes today. These "soundings" are embedded as part of a layering of presence, a continuing "being" in the world, and a form of expanded reality in these landscape settings. It is a normal ethnographic sensorium of an "afterlife consciousness" that has become embedded and imprinted on these battlefield spaces. This has created a "hauntscape" of multiple immiscible social and mental "fields" that <u>potentially</u> can only be accessed, on a <u>consistent</u> basis, through cultural resonance. This assumption implies **"the existence of an essential and fundamental relationship between ghosts, mind, and consciousness" (Beichler 2011:30).**

Finally, one may ask why these "ghosts", if they exist, remain on the battlefield where they died on September 17, 1862. There are several possibilities. I will only mention (briefly) a few here. These include:

- Military orders; and
- The concept of the "Good Death".

Standard military orders are significant, but often overlooked in a battlefield "apparitional experience", especially the following:

- "To quite my post when properly relieved";
- "To be especially watchful at night"; and
- "To talk to no one except in the line of duty".

Do our "ghost excavation" performance practices at night, and contextual to "identities" affiliated with "commanding officers", allow for the manifestation of an "afterlife consciousness" because soldiers, those who remain attached to duty, honor, and commitment, communicate with us in the "line of duty"? Do these soldiers remain because they were not accorded the rites and performances of the "Good Death"?

According to Clinton (2009), **"the deathbed of a loved one was perhaps the most hallowed of Nineteenth Century ritual settings (2009:4).** The "Good Death" was a <u>prepared</u> death, surrounded by family at home, and a subsequent burial in the family plot. The American Civil War battlefield changed that. The ritual of the "Good Death" was never completed, in many instances, for the soldiers who fought and died at Burnside Bridge. Some of these soldiers remain buried where they fell (mostly Confederate). Many others are buried in the cemetery at Sharpsburg, Maryland. These soldiers never went home!

Photo 21: The Civil War Cemetery at Sharpsburg, Maryland

Are the audio manifestations that we have recorded at Burnside Bridge the "afterlife consciousness" of those soldiers who fought, died, and are buried elsewhere other than "home"? Are these the "ghost soldiers" who remain on duty, are vigilant at night, only communicate to individuals that they identify as their comrades and/or officers, and who never experienced the "Good Death"? This is the theory that I am currently working with, as we continue to investigate the manifestations of past presence, as an "afterlife consciousness" of the "culture of war" at Burnside Bridge, and other battlefields. The learning process continues with the "ghosts of the battlefield" as apt "pupils" and "instructors" for us who investigate their continuing presences!

Haunted Prisons:

Learning the Gaze of Isolation and Confinement

I once "entertained" the belief that "extreme" investigative presence was suitable for exploring haunted prison space. This was in 2006-2008, long before "paranormal reality TV" popularized it. "Taunting", as a form of control, was acceptable fieldwork behavior, I thought, because it was contextual to what occurred in the past.

I believed, then, that "ghosts", or a "ghostly" presence, could occur by some act of intrusion or disrespect in their space. This was seen as a form of "provocative hunting" and was thought to be effective in "unearthing" a past presence. This occurred because I believed, at that time, that the contemporary emotional "charge" or frequency ("provoking") resonated with a particular past, highly-charged situation in haunted space.

Today, I learned that this is a mistaken notion of responsible (and moral) investigative fieldwork. We must learn to alter our "treatment" of "prison ghosts", and other "institutionalized" hauntings. It's a matter of principle. This is a re-consideration of politics, control, and authority from what occurred in the past.

The "frozen" time of haunted space are places within space that contain repressed mini-moments (sometimes seconds) of a "life" time. These liminal spaces illuminate a period between architectural and human time, a "ghost" time. This "ghost time" echoes human dreams, anxieties, purpose, struggle, and completion without borders, corridors, and exits.

All explorations of haunted space involve a type of archaeology, the "unearthing" of a fragmented narrative that traces a moment in human time. Its exploration is an entry, a portal into more that past spatial dimensions that are measured as a contemporary ambience of deviation and anomaly. What remains are human vestiges of cultural expressions and social control. This becomes an important lesson we must learn in certain institutional frames, such as prisons.

An "extreme investigative presence" continues the mentality of politics, and complete control and authority into the "afterlife" of the "ghosts" that haunt these locations. Is that morally correct, or being socially responsible? Haven't they already "served" their "life" sentence? Isn't this a form of mental cruelty, a psychological torture that was already experienced during life? Why continue this "life sentence" into the "afterlife"?

Photo 22: Guilty! The "taunting" shower scenario in Cell Block 12 at Eastern State Penitentiary, Philadelphia (Pa.).

Photo 23: "Guilty"! The "Al Capone" scenario at Eastern State

We must end the concept of a permanent haunting domicile for criminals who deserve to remain there! Are they "guilty" of an "afterlife" sentence because of their past criminal acts, or their "guilt"? How do we learn the "truth"? What happens to the "truth" when the prison becomes a contemporary Halloween attraction as well (as it does at Eastern State)? How does this "play" to those who may haunt these spaces?

What are the cultural configurations of the prison complex, as "institutionalization", and how does it affect our learning the "truth"? How is Alcatraz Island, with tours that focus of history, different from Eastern State Penitentiary, with its focus on a haunting ambience, and how does this affect the true nature of what really haunts these places?

Is "anything goes", a legitimate form of constructive fieldwork because these prisoners have been constituted as "civilly dead"? A "civil death" is an individual who has been stripped of all civil rights. Is this the proper way to learn about a haunting, or even the possibility of "ghosts" haunting these prison institutions?

Photo 24: Eastern State Today

Photo 25: Another View of Eastern State

Photo 26: Alcatraz Island Today

Photo 27: Another View of Alcatraz

A prisoner is a perfect example of an individual in a liminal position (betwixt and between), even <u>before</u> death (and possibly between the world of the living/dead as "ghost"). The individual is:

"a divided figure: a redeemable soul, but also an offending body; a citizen-in-training, but also an exile from civil society....." (Smith 2008:248).

Within the prison walls, these "inmates" appear to be "perfect" candidates to "exploit" as potential "ghosts". The architecture, dress, food – all of it – were planned to play a proactive role in accomplishing the prison's institutional goals and purposes. These goals and purposes not only create an atmosphere for potential "ghostly" sightings, they serve as a "stage" in which to conduct investigative performances related to those haunting uncertainties.

Many of the spaces at Eastern State in Philadelphia appear today as haunting aspects of the "afterlife" of the prison. These spaces are perceived to be "haunted" by the cruelties of past activity and occupation that reside in antiquated machines and furniture, functionally-assigned rooms, and "artifacts" of acts that are now abandoned throughout the site. This leads some to conclude: "if there ever were "ghosts", they surely would haunt here"!

Photo 28: Inside the Walls at Eastern State

But how are "clues" to obscured events, found in the debris of image and audio, and measurement and experience, to be decoded, and which reflect an accurate portrayal of "what happened here" and who continues to remain incarcerated here? How can we link particular spaces to specific "ghosts"?

How do we justify the raw experience of confronting the physicality of a now discarded infrastructure in ruin, and the abandoned system of beliefs and actions that were followed and enacted here? How do we document the continuation of past function and purposeful acts today as a "haunting"? Is something still unresolved? Did "someone" from past history manifest <u>intentionally</u> as an <u>intelligent</u> act to protest their treatment or confinement, or to establish/reinforce "political" order and institutional discipline?

Fieldwork at Eastern State demands more than a "ghost hunt". It demands a work of socio-historical research and empathy, looking into archives and testimonies, and a personal interaction with the layers of presence that parallel the various stages of ruined facilities and prison space. It must call attention to the horrors of institutional philosophy and the dehumanizing strategies of acts that are encoded within this institution. Have they become "imprinted" on these prison walls?

The fieldwork that should (must) be enacted is surely not a form of "entertainment" or a haphazard search, as this further adds to the layering of presence that is already here. We must separate the "ghosts" from the "ghosts of place": radiating bandwidths, cell phones, institutional static and chatter, pressure changes, and ultrasounds (to name a few). These are acoustic atmospheres. Some are thick with memories and "imaginings". Others may be purposeful acts of "absence", reflecting the institutional policy of utter "silence", an important sensual component of Eastern State's initial 75 years of penal occupation.

Along these cell blocks, which are the "imprints" and "who" are the interactive presences? Are they the administrative personnel, the "visitors", or those who were incarcerated here? All of this background can (must) be used in fieldwork "excavating" performances, with an emphasis on the institutional power of social control. The fieldwork can be divided into determining the effects of:

- The physical experience of confinement; and
- The situational nature of the confinement inside the prison walls.

These can relate to:

- "House" (Institution) rules and structured humiliations (the use of "hoods" for interior movement; the loss of identity);
- "Coping" (situational withdrawal; refusing to cooperate; "messing-up"; how to manipulate the system); and
- The inmate sub-culture (how to communicate without talking; the mobility and flow within the walls).

As investigators, we must "tune-in" to the material signatures of former life and control at Eastern State. We must constantly ask ourselves how much more we can learn there. In the words of Catherine Spooner (2010), a senior lecturer in English Literature:

"ghosts, with their tendency to unsettle our comfort zones, mitigate against complacency,

and encourage us to keep asking questions"
(2010:183).

We must incorporate a holistic approach to the study of haunted prison culture, not a simple measured scan, or a quick "ghost hunt" habitualized EVP session. We must fully explore what individuals (guard, inmate, administrative personnel, guest, or family) did in each space and situation within the prison. We must explore the possibilities of "when" and "how" they did it, with "what", "whom" and "why", and what they did not do (or could not do) and "why".

This fieldwork becomes, not a re-enactment or re-construction, but a <u>construction</u> of certain scenarios framed as hypotheses that can help us to understand any haunting phenomenon that is encountered within the prison. This we do by testing specific acts and events that occurred (may have occurred) in the prison.

Much of this is <u>not</u> new. It can be seen occasionally on TV. I have done it for more than four decades now, though not always at haunted prisons. Still, the majority of "ghost hunts" or paranormal TV does <u>not</u> incorporate it into a <u>seen</u> methodology in their programming. There is no need. After all, this is viewed as "entertainment", not science.

The "prison culture" at Eastern State, for a large part of its history, was a "culture of silence". This is an important investigative consideration. What we can learn from that "silence" should be approached with caution. Blesser and Salter (2009), in their book on aural architecture, make the following observation:

"Silence creates large acoustic arenas as a common resource....(and) **only the highest-quality acoustic arenas, with very low background noise, communicate silence (2009:32).**

Eastern State Penitentiary today does <u>not</u> conform to a high quality acoustic arena which communicates silence. There is continuous "outside" background noise that occurs throughout the prison. These noises are not "paranormal" in nature, nor are they the sonic broadcasts of former inmates, guards, or administrative personnel. They are due to the decay and neglect of an institution in ruin.

There are also the contemporary auditory elements of historic exhibits, "ghost hunts", and a seasonal "haunted house" that <u>do</u> create additional aural elements as residual auditory presences. Any EVP session that sounds like "voices" from the past, even in this prison culture of silence, may be merely recording these background noises.

This means that the prison soundscape can become **"severly degraded when the aural environment falls victim to intruding noise (Blesser and Salter 2009:32).** In most cases, EVP recordings, especially those recorded in the cell blocks that were in use during this "silent" era, are probably <u>not</u> the "voices" of the dead.

The "prison culture" at Eastern State was controlled with a purpose. The administrative body that governed there demonstrated this policy by controlling the communicative expressions within the prison setting. Adam Jaworski, an expert on aural silence, calls this control **"the politics of silence and the silence of politics" (1993).** A "ghost hunt" here should not forget the importance of this "politics of silence". Its importance lies in the fact that:

"when an acoustic space is repeatedly used for a specific purpose over a long period (as Eastern State was during the era of "silence", 1831-1903), **the culture begins to associate its aural personality with that silence" (Blesser and Salter 2009: 362-363).**

This association led to a "tradition" at Eastern State of "silence", extending possibly after death, continued by the "ghosts" (both inmate and guard) who may still haunt (and inhabit) the cells and cell blocks of the former prison.

The architecture of Eastern State, and its contemporary physical spaces as "ruin", is transforming the dynamics of haunted space. And now, the "ghosts" of former inmates remain "free" to change their arenas, and manifest in other former "prohibited" spaces. Or do (can) they do so? Does their former social behavior (and the "prison culture") continue to "prohibit" this freedom of movement?

This can only be observed and/or recorded when the investigative team is correctly "targeting" activities with locations, and using specific behavioral patterns that resonate with the "ghosts". This process can only be learned through extensive historical and ethnographic research into the dynamics of the prison's policies and their execution. The "ghosts" of Eastern State are still waiting to be released from their confinement. After all, it may still be a <u>functional</u> prison to those that remain there!

Photo 29: The Interior of Eastern State Penitentiary

Haunted Centers of Learning

"So now then we begin again this history of us"

- **Gertrude Stein, "The Making of Americans (1925).**

This Happy Breed (1944), a film by David Lean, is a subtle tale of what haunts us at home in our own familiar surroundings. It is not about "demons" or other supernatural beings. It is not even "paranormal". It is "us" who haunt our houses. We are the "ghosts" of these haunted houses. There is one scene in which the occupants of a house wonder if their lives will become imprinted onto the walls, and if others who come afterward, **"will feel any bits of us hanging round the place".**

Children, frequent inhabitants of our homes, play a significant role as a "mediator", between the worlds of the living and the dead, and sometimes a "haunting" in the home is a re-acquaintance with a "child-like haunting".

Gilles DeLeuze has observed:

"There is no present which is not haunted by a past and future, by a past which is not reducible to a former present, by a future which does not consist of a present to come" (quoted in Gilbert 2004:88).

What better place to learn about these three transformations: "us" as "ghosts", the "children's hour" (as a particular form of "haunting"), and the juxtapositions of time than a museum. Little did we know when we began our investigation of the Brunswick Railroad Museum (Brunswick, Md.) that we would learn first-hand about these transformations in a haunting field that encloses the museum's walls and its architectural stories and contemporary exhibits from the past.

We also learned that the museum is more about a childhood that haunts, than a history of the "Red Men" who occupied the building, the plays and concerts enacted there, the bar where members enjoyed a drink, or even the history of railroading in Brunswick. Though all are potential sources of uncertain residual agency and energies, it is the children who really haunt the museum (at least it was for us).

Photo 30: The Brunswick Railroad Museum (Brunswick, Md.)

Photo 31: The Brunswick Railroad Museum

In **How Buildings Learn: What Happens After Their Built (1994),** Stewart Brand explores how buildings "live" in time, and how different layers of function and occupation are changed to accommodate new uses. As Henri Lefebvre, in **The Production of Space,** states:

"No space ever vanishes utterly, leaving no trace".

A museum symbolizes that concept, and a "haunted" museum transforms that symbolization. The Brunswick Railroad Museum is an example of that symbolic transformation. But it is not the building that is haunted so much as it is the exhibits that are contained there.

At the Brunswick Railroad Museum, memories and their attachments to certain objects, images, and documents plays an important "ghosting" role in the mise-en-scene of potential haunting phenomenon. But are the "ghosts" imprinted with the contents of the displays and events in time, or with actual presences who associate themselves with the objects on display? We came here to learn and investigate whether the haunting is an "imprint", a "haunted collection", or a "live" presence from the past.

Photo 32: The Discussion with Investigators Before the Fieldwork Began

We began with a peripatetic walk through the 2nd and 3rd floors to get a sense of the spaces and displays, and record the interior soundscape on these floors.

Photo 33: The Peripatetic Walk

Photo 34: The Peripatetic Walk

During the walk on the 2nd and 3rd floors, we did sense a presence in the hallway leading to the entrance to the "Red Men" Lodge on the 3rd floor. This presence was also felt while we were enacting various scenarios on this floor. And we later recorded footsteps as we descended down the stairs to the 2nd floor.

Photo 35: The Museum Stairs

There was also a sensed presence on the 1st floor, in the downstairs bathroom area, below the steps leading to the 2nd floor. One of our investigators felt that this presence was "aggressive", and she said that she felt being touched in this area several times. We do not, however, have verification of these tactile sensations.

Photo 36: The Area of "Presence" on the 1ˢᵗ Floor

In our exploration of the museum's potential haunting uncertainty, we began with the building's functional layering (<u>before</u> it became a museum). This exploratory "excavation" included enacting contextual (and resonating) scenarios that were relative to the social experiences and cultural situations that would have occurred during particular functional phases of the building's prior histories.

A haunting can be relative to the "character" of a place. This "character" can derive from the social experiences that are enacted there. These social experiences are **"mediated upon the landscape** (or building) **upon which that experience unfolds" (Bell 1997:831).** We had a number of social experiences in mind for our "excavating" performances.

On the 3rd floor, the location of the International Order of Red Men (I.O.R.M.), Delaware No. 413, we conducted several "excavations". This area was considered the "wigwam of the tribe". The I.O.R.M. was a fraternal organization whose goal was **"perpetuating the beautiful legends and traditions of a vanishing race and keeping alive its customs, ceremonies, and philosophies".** There is a cornerstone that is carved on the side of the building of the current Railroad Museum. It reads: **"G.S.D.-413".** This signifies "Great Sun Discovery" and 413 years after the year of discovery by Columbus (1492). This dates the building, and its occupation by the "Red Men", to 1905 (1492+413= 1905). We used this historical information in the performance of various scenarios on the 3rd floor.

We also enacted a number of other scenarios on the 3rd floor in order to "unearth" any continuing "Red Men" presence. These included:

- The use of a secret "password" to gain access to the 3rd floor;
- The use of the "Red Men" slogan: **We visit the sick, bury the dead, educate the orphan";**
- The **"kindling the Council fire"** (opening a meeting);

- The initiation ceremony of a **"Paleface"** (a non-member); and
- Activities with the **"Degree of Pocahontas"** (the Ladies Auxiliary).

Photo 37: Members of the I.O.R.M. in Full Regalia

We also used a copy of the "Red Men Diploma and Legendary and Historical Chart" and an "I.O.R.M. Record", as "trigger" mechanisms, placing them on the stage (and marking their location). These <u>contextual</u> objects did not move during the enactment of our scenarios.

Photo 38: The "Red Men Diploma"

Photo 39: The "I.O.R.M. Record"

In the spaces where the men **"ride the goat",** and relative to the function and social experience of the "Red Men" organization, we did not record or document any <u>solid</u> confirmations of "apparitional experiences". Was this a function of <u>contemporary</u> layering (the "model train" exhibit) and the "tourist gaze" (visitors viewing the exhibit, and unconscious of the previous function and history of the 3rd floor)? Did this contemporary "layering" function to "erase" prior uncertain events or acts? We do not know at this time. But if they did (and continue to do so), it would change our concept of "residual" presence and energy.

This means we would have to re-learn the reality of a "residual haunting", or expand its dynamics to include variables of frequency, strength of presence, and other factors. This would involve developing a stratigraphy of occupation and presence for haunted space. And it must include a matrix of associations that may alter the manifestations of prior haunting "residual" elements.

If the present superimposes itself onto a space, especially if it is non-contextual to what occurred in the past, then this contemporary presence would impede what presences were "recorded" earlier from a time more remote in the past. The contemporary past would suppress the historical past. In the case of the 3rd floor at the Railroad Museum, the presences of "Red Men" occupation, if this theory is correct, would be suppressed by the non-contextual presence of the "model train" exhibit and the people who come and see that particular display. This process of possible haunting dynamics should be explored more thoroughly at the museum.

Next, we enacted scenarios on the 2nd floor, the location of the historical exhibits. These were contextual to the history of the town (as a railroad hub), and the history of the surrounding area. The exhibits that were in place were used as "triggers" to resonate with this history, and several cultural scenarios that resonated with them. We also used contextual "trigger" words in our conversational dialogues, and acoustical "triggers", such as railroad sounds, as particular "soundmarks" to "unearth" past memory and any "afterlife consciousness" that may remain here, or had become "attached" to any of the exhibits.

Photo 40: The 2nd Floor "Excavation"

Photo 41: The 2nd Floor "Excavation"

We used vocal expressions from the 1890's, as elements of an "ethnography of communication" (E.O.C.) relative to that historical period. The E.O.C., first developed by socio-linguist Dell Hymes, is based on socio-cultural context. It is believed that this context might shape culturally-specific linguistic interaction, what "ghost hunters" call "EVP". Certain phrases were used instead of the usual "ghost hunting lingo", such as "Is anyone here with us tonight", or the equally non-culturally-specific phrase, "Can you show us a sign of your presence".

The purpose of a "ghost excavation" is to become "identified", by any interactive past presence that might linger, as an individual from a particular time period **and** cultural horizon. If there is any "interactive" presence in a particular space, there is a higher probability that they would interact with a <u>recognizable</u> human being (a <u>participant</u> in that culture) rather than an "outsider", one who is <u>not</u> recognizable because they use the wrong <u>communicative</u> code (among other non-contextual elements: clothing, electronic devices, and "watch and wait" non-interactions).

Some of these communicative "triggers" included expressions from the 1890's, such as:

- **"Tell me pretty maiden, has your mother got any more like you?"**
- **"cheese it!" ; and**
- **"no flies on him".**

Some of the scenarios that we enacted on the 2nd floor, in relation to its former function as a dance floor, community center, and bar included the following:

- The announcement of an "accident" at the railroad yard in front of the "medical" exhibit;
- The wicker body-basket scene for an injured railroad worker, enacted with an actual wicker body-basket (in the vicinity); and

- Various barroom scenes, at the location of the former bar. These included general conversations among the patrons, a discussion of other places to drink locally (**"No drink here. Let's go down the road to Knoxville!"**), and an exchange between the male patrons and some females.

Most of these scenarios proved uneventful, though we did hear some sounds around the "bar". During the "accident" scenario, we did hear sounds of "whimpering" (not recorded). One of the investigators also perceived the presence of a "Jenny" whom she felt was "screaming terribly" (not recorded). Another sensed presence was "Carmen" or a "car man".

We also used some non-contextual acts to elicit some kind of "negative" response or reaction. These included:

- Women portraying "Red Men" prospective candidates. There were no female members among the "Red Men"; and
- Non-members attempting to gain access to the 3rd floor, and the "wigwam".

No <u>documented</u> activity was noted during these investigative performances. Was this because they were non-contextual, and did not resonate? Or, was it because there is no haunting presence that is associated with these scenes of uncertainty? We don't know the answers, yet.

It was at the "school" exhibit, with several investigators portraying "schoolmarms " that proved to be the most active area during the "excavation". Here, we recorded a number of E.O.C. (or EVP). They can be heard on our website at www.ghostexcavation.com. At the school exhibit, there was a sensed presence of a "Margaret" and/or "Gloria", but whether these are "students" or "teachers" is not known. While we were filming these scenarios at the "school" exhibit, the video feed kept going out of focus, as if someone was walking in front of the video camera.

In the display, there were school-related items, such as books, desks, and a Ben Franklin stove. There was an enlarged poster of "school children". The "school scenario" initiated with the "Pledge of Allegiance", and continued with other school-related activities (such as particular "lessons" on arithmetic, geography, and reading). When Mary Becker, as one of the "teachers" introduced herself, there was an aural response that said **"I don't like Miss Mary".**

Photo 42: The "School" Exhibit

The theme of childhood is a frequent trope in films about a haunting, especially in relation to memories of fear or loss. In the film, **Night at the Museum (2006)**, childhood anxieties are exploited regarding the uncertainty of "dead things" that can re-animate after closing time. The uncanny character of a museum was used to produce the anxiety of encountering the stored energies of the past. In much the same way, the paranormal reality TV program, **Haunted Collector**, exploits the psychometric nature of "haunted objects". Another example of this is the classic movie, **Citizen Kane (1941)**, in which a man ("Kane") dies trying to re-animate a lost world. "Rosebud" is the "ghost" of an undead childhood that haunts "Kane" throughout the film.

Are the "ghosts" of children at the "school" exhibit in the Brunswick Railroad Museum "active" ghost children, or the "residual" presence of a childhood memory embedded in the display? Are memories of childhood, rather than "active" presence, the sonic manifestations that "haunted" us during the "excavation"?

If, as Adam Phillips (1999) suggests, adulthood is the "afterlife" of childhood, are we ourselves the "triggers" that enabled "ghostly" children to manifest, rather than the objects of the exhibit itself? Did we become their "teachers", or were they the "instructors" in the "afterlife" existence of historical "childhood" presence at the museum?

The E.O.C. that we recorded, with Mary Becker as the "teacher", might shed some light on the answer. This would seem to confirm an "active" rather than a "residual" presence. You can hear these responses at www.ghostexcavation.com. It appears that these aural manifestations at the "school" display do "exhibit" a teaching tool that instructs us about "presence" and "who" remains from the past at the Brunswick Railroad Museum.

One question remains: Why are there "ghost children" at the Brunswick Railroad Museum, and little indications of adult "ghosts"? Perhaps, children return to a "happy place" or a good (positive) memory of a childhood experience, as symbolized by the school exhibit. The adults do not "return" because perhaps the location (the museum building) was merely someplace to get away from "life" and its problems and complications, and not a place to remember or have fond memories of.

If this hypothesis is correct, would this alter the belief and perception of why some locations have an "afterlife" presence and others do not? Would it explain why some children haunt a particular space, and adults do not? Or perhaps, some adults return to a place of childhood innocence, rather than a site of adult occupation with all the "baggage" that this entails?

Is a haunting in some cases, such as the Brunswick Railroad Museum, the "afterlife of childhood" of adults as well? And is that the reason why their absence (as adults) is not present there? This question beckons us: are these manifestations that we recorded the "ghostly" presence of "children", or the presence of "adults" reliving their childhood memories? A need for future "excavations" is required to explore these uncertainties.

What is paradoxical about the Brunswick Museum haunting is that adults play the inevitable prominent role in children's lives, and the "heritage of childhood" (such as the school exhibit) is generally undertaken by adults. Yet it's the children who are "active" creators of their own cultures (cf. Darian-Smith & Pascoe 2013).

The haunting of a cultural heritage symbol of childhood (the school exhibit) alters this normal expectation of influence, usually revolving around the adult world which defines how children, and this particular period of childhood, is defined and understood. With the exhibits dominated by an adult world, it's the uncertainty of a continuing "childhood" experience, rather than an adult presence, that may haunt the museum's 2nd floor. Even in ghost research, though "ghost children" have been reported in many a haunted space, there is little mention of a "child haunting" based on the cultural heritage of children, a child's "ghost culture".

A child is not a passive being, unspoken and vulnerable, needing constant adult supervision. As Nick Lee suggests, researchers need **"to see children as human beings, active in social life, rather than as human becomings, passive recipients of socialization" (2001:49).** If they remain as "ghosts", they should be treated accordingly, as "active participants", not passive entities who are gently persuaded into doing something for the video camera, or saying something for the audio recorder.

This is a view of a "child ghost" as <u>independent</u>, not necessarily feeling alone and afraid, and waiting for adult supervision. This <u>independence</u> is what occurred at the "Knick" when Mary Becker was playing a "counting game" with a "child". The child <u>independently</u> played "her" (?) own numbers game, by "counting" ahead of Mary's count at times, or saying (continually) the number "5" while Mary was counting. The child showed no fear or a need for supervised play. You can hear this at www.ghostexcavation.com.

Ultimately, institutions of learning, or museum exhibits about learning, are the most suitable locations to <u>learn</u> about haunting phenomenon. They contain transformative spaces, consisting of an educational venue that appears to continue intellectual stimulation and simulation. We cannot find a more meaningful location which "haunts" us into learning more about them.

"A thing which cannot be understood inevitably re-appears like an unlaid ghost" (Sigmund Freud).

Let's end the "life" of a "ghost" as a paranormal anomaly, and make them "human" again through our research and fieldwork. And more importantly, let's learn from that "apparitional experience"!

Post-Script:

A Future Lesson Plan(?)

What could be more challenging, more instructive than the "traditional" haunted house, one that is "haunted" not by "who" lived there, or what occurred there, but the very house itself! This is the sentient, animated house. This will be the next agenda we will explore of haunted space and its "excavation", as we go back to a future manifestation of "presence".

Though common in popular culture in various genres, from Poe's "The Fall of the House of Usher", and Hawthorne's, "The House of the Seven Gables" to Shirley Jackson's book as movie, "The Haunting" and the TV horror classic, "American Horror Story (Season 1)", the "house" has manifested many times through the years.

Why has this popularity of the haunted house trope (in movies and on TV recently) become increasingly associated with TV viewings in the home (see the movies, **Poltergeist, The Ring**, and **Ring 2,** among others). Does this allow us to "forget" threats associated with more remote (and "traditionally" haunted) parts of the house, such as the attic or basement? Does the TV, as a "receiving" device aid in these manifestations? And how does this relate to the house, rather than people or "haunted objects", being responsible for a particular haunting, or even a "haunting vortex"?

How can one investigate such a house whose spaces are "popularized" as quite "unliveable"? Does such a house exist independent of the imagination? Who is really in "possession" of such a house, when not even the "ghosts" are welcome or "expectant"? This is a house that would truly educate us in many ways. Perhaps, next time.......?

Bibliography

Barba, Eugenio (1986). *Beyond the Floating Islands.* New Jersey: PAJ Publications.

Batzler, Louis R. (2011). "Panel: Ghosts: Mind and or Matter??" Academy of Spirituality and Paranormal Studies, Inc. 2011 Annual Conference Proceedings. Bloomfield, Connecticut pp. 68-69.

Beichler, James E. (2011). "Apparitions R' Us: Far More Than Just a Ghost of a Chance". Academy of Spirituality and Paranormal Studies, Inc. 2011 Annual Conference Proceedings. Bloomfield, Connecticut. pp. 24-38.

Bell, M. (1997). "The Ghosts of Place". *Theory and Society.* 26: 813-36.

Blesser, Barry and Linda-Ruth Salter (2009). *Spaces Speak: Are You Listening? Experiencing Aural Architecture.* MIT Press.

Bourdieu, P. (1977). *Outline of a Theory of Practice.* Cambridge: Cambridge University Press.

Brant, Stewart (1994). *How Buildings Learn: What Happens After They're Built.* Penguin Books.

Braud, W. and R. Anderson (1998). *Transpersonal Research Methods for the Social Sciences: Honoring Human Experience.* London: Sage.

Darian-Smith, K. and C. Pascoe (2013). *Children, Childhood, and Cultural Heritage.* New York: Routledge.

Delaney, Carol (1988). *Investigating Culture: An Experiential Introduction.* Wiley-Blackwell.

Fischer-Lichte, E. (2008). *The Transformative Power of Performance: A New Aesthetics.* S. Jain (Translation). London: Routledge.

Fontana, David (2009). *Life Beyond Death: What Should We Expect?* London: Watkins.

Fraser, John (2010). *Ghost-Hunting: A Survivor's Guide.* History Press.

Geertz, Clifford (1973). *The Interpretation of Cultures.* New York: Basic Books.

Germain, Gil (2009). *Spirits in the Material World: The Challenge of Technology.* Plymouth, U.K. : Lexington Books.

Gilbert, Ryan (2004). *Groundhog Day.* London.

Goldman, Marcio (2003). "Os Tambores Dos Mortos os Tambores dos Vivos: Ethnografia, Antropologia, E Politica en Ilheus, Bahia". Revista de Antropologia 46(2):446-476.

Gordillo, Gaston (2004). *Landscapes of Devils: Tensions of Place and Memory in the Argentinean Chaco.* Durham, North Carolina: Duke University Press.

Hendon, Julia A. (2010). *Houses in a Landscape: Memory and Everyday Life in Mesoamerica.* Durham, North Carolina: Duke University Press.

Heidegger, M. (1971). "Building, Dwelling, Thinking" in A. Hofstadter (Translation) *Poetry, Language, Thought.* London: Harper and Row. pp. 143-161.

Hodder, Ian (1999). *The Archaeological Process: An Introduction.* Oxford: Blackwell.

Ingold, Tim. *What is an Animal?* London: Routledge.

Jaworski, Norman (1993). *The Power of Silence: Social and Pragmatic Perspectives.* Newbury Park, California: Sage.

Jordania, Joseph (2011). *Why Do People Sing: Music in Human Evolution.* Logos.

Lee, Nick (2001). *Childhood and Society: Growing Up in an Age of Uncertainty.* Philadelphia: Open University Press.

Noory, George and Rosemary E. Guilley (2011). *Talking to the Dead.* New York: Tom Doherty Associates Books.

O'Donnell, Elliot (1969). *Casebook of Ghosts.* (H. Ludlam, Editor). New York: Taplinger Publishing Company.

O'Reilly, Karen (2005). *Ethnographic Methods.* London:Routledge.

Palsson, Gisli (1994). "Introduction: Beyond Boundaries" in *Beyond Boundaries: Understanding, Translation, and Anthropological Discourse.* (G. Palsson, Editor). Oxford: Berg. pp. 1-40.

Pearson, Mike and Michael Shanks (2001). *Theatre/Archaeology.* London: Routledge.

Phillips, Adam (1999). *The Beast in the Nursery.* London.

Pink, Sarah (2007). *Doing Sensory Ethnography.* Los Angeles: Sage.

Roundtree, David (2010). *Paranormal Technology.* Bloomington, Indiana: IUniverse.

Sabol, John G. Jr. (2007). *Ghost Excavator.* Bloomington, Indiana" AuthorHouse.

(2008). *Battlefield Hauntscape.* Bloomington, Indiana: AuthorHouse.

(2009). *The Politics of Presence.* Bloomington, Indiana: AuthorHouse.

(2011). *Digging Up Ghosts.* Brunswick, Maryland: Ghost Excavation Books, Inc.

(2013). *Burnside Bridge.* Brunswick, Maryland: Ghost Excavation Books, Inc.

Schneider, Rebecca (2011). *Performing Remains: Art and War in Times of Theatrical Reenactment.* London: Routledge.

Shanks, Michael (2012). *The Archaeological Imagination.* Walnut Creek, California: Left Coast Press.

Sheldrake, Rupert (1999). *Dogs That Know When Their Owners Are Coming Home.* London: Hutchinson.

(2012). *The Science Delusion: Freeing the Spirit of Enquiry.* London: Coronet.

Smith, Caleb (2008). "Detention Without Subjects: Prison and the Poetics of Living Death". *Texas Studies in Literature and Language.* 50 (3), Fall. pp.

Spooner, Catherine (2007). *Contemporary Gothic.* Reaktion Books.

Toop, David (2010). *Sinister Resonance: The Mediumship of the Listener.* Continuum Publishing Group.

Underwood, Peter (1983). *No Common Task: The Autobiography of a Ghost Hunter.* London: Harrap Liwatec.

(1994). *Nights in Haunted Houses.* London: Headline Book Publications.

Waylin, Casper (2012). *How To Research Haunted Locations.* On-Line E-Book from Amazon.com.

Wikan, U. (1992). "Beyond the Word: The Power of Resonance". *American Ethnologist* 3: 460-482.

Wilber, K. (1983). *Eye to Eye: The Quest for the New Paradigm.* Boston: Shambhala.

Biography

John Sabol is an archaeologist, cultural anthropologist, actor, and author. As an archaeologist, he has documented and recorded the manifestations of past soundscapes at haunted ruins. As an actor, he has appeared in many movies, TV series, and educational TV programming, including the Sci-Fi classic, **Dune (1984)**, and the A&E TV series, **Paranormal State.** He has written 16 books on his fieldwork, methodology, and his personal experiences on location filming, and his work at haunted ruins around the world. He has been a frequent guest on numerous radio and internet talk shows, among them, Beyond the Edge Radio, The Paranormal View, Para X Radio, Blog Talk Radio, The Grand Dark Conspiracy, and Rusty O'Nhiall's "Mysterious and Unexplained" on PsiFM (Australia).He has also worked on international educational documentaries (in Spain).

He is the director of several documentaries that are accounts of immersions into past ethnographic soundscapes at historic sites now in ruin. In these ruins, he has recorded manifestations of past cultural behavioral fields, including the "culture of war" on several Civil War battlefields and, most recently, coal-mining cultural vestiges and traces of past presence at Centralia, Pennsylvania. He has organized (and played a role in) theatrical "stagings" and "ghostings" at haunted locations which recorded "spiritscape" sites at these haunted locations.

He has developed numerous scripts and storyboards for these documentaries, as part of a "ghost excavation" series of mediated venues. He has presented video clips and audio tracks of these documentaries at various scientific conferences and popular culture expositions in Europe, Canada, and the USA. Upcoming projects include writing, directing, and acting in a documentary about Civil War hauntings along the Mississippi River, and documenting a series of Sasquatch habitation sites in the Southeast U.S.

Future speaking appearances include T.A.G.(Theoretical Archaeology Group) Conference at the University of Chicago, Oriental Institute, The Hindsdale Project in Hindsdale, New York, Popular Cultural Conferences in Washington, D.C. and in Niagara Falls, Ontario, Canada, the G.H.O.S.T.S. Conference in Ontario, Canada, several conferences in England, and a major symposium and conference in Edinburgh, Scotland in 2014. At the Edinburgh Conference, he will also direct a "ghost excavation" at the legendary and very haunted Greyfriar's Cemetery.

He can be reached via email at cuicospirit@hotmail.com. His website is: www.ghostexcavation.com. and he can be found on Facebook ("Ghost Excavations with John Sabol" and "Beyond the Paranormal").